Date Due

D0850615

HIGH-ENERGY
ASTROPHYSICS

HIGH-ENERGY
ASTROPHYSICS

TREVOR C. WEEKES

CHAPMAN AND HALL LIMITED

11 NEW FETTER LANE, LONDON EC4

First Published 1969
© *Trevor C. Weekes 1969*
Printed in Great Britain by
Butler & Tanner Ltd, Frome and London

Distributed in USA by
Barnes & Noble Inc.

JOINT UNIVERSITY
LIBRARIES
NASHVILLE TENN.

Observatory
QB
464
.W4

645992

To my parents

Contents

List of Plates

Acknowledgements

The monograph was written at the suggestion of Professor N. A. Porter who did much to kindle the author's interest in this subject and whose help at all stages in the preparation of this work is gratefully acknowledged. The text is based on a series of lectures to senior students in the Physics Dept., University College, Dublin, in 1966. I am grateful to Dr P. K. MacKeown for reading the chapter on neutrinos and to Dr C. D. Long and G. H. Rieke for reading the entire manuscript. Those mistakes that may still be present are the sole responsibility of the author.

T. C. W.

Preface

It would be pretentious for any work of this length to claim to be more than an introduction to but one branch of astrophysics. The scale of astrophysics is so immense, the range so diverse, that there are few physical processes not represented in it to some degree or other. To give even a brief outline of all astrophysical processes in this short space would lead to a mere listing; on the other hand to give a comprehensive treatment of just one topic would demand a greater background of astronomy than the physicist normally has. The subject matter here has been restricted to high-energy astrophysics, a subdivision that includes many of the most interesting areas of research at this time. The topics that have been included come under the high-energy label in a number of ways: high energies relative to the rest mass of the object (supernovae, radio galaxies, quasars), individual quanta possessing high energies (cosmic rays, x-rays, gamma-rays), possible large cosmic energy densities (neutrinos, microwaves). The list is by no means complete and inevitably reflects a subjective bias on the part of the author.

Although many astrophysical theories have been developed in detail, most of the presentation here will be in outline only. The details and mathematics are generally omitted, since it is felt that with the present controversy that exists in most branches of the subject, it is the lines of thought that are most important. Where possible the observational aspects will be emphasized, since astrophysics suffers from a dearth of meaningful observations; the extension of existing techniques is thus all important for the development of the discipline.

A word of warning is necessary for those accustomed to the hard
and fast laws of laboratory physics; there are few absolutes in
astrophysics. At this moment the subject is very much in a state
of flux, with almost complete revolutions every decade. In most
cases theoretical speculation is in terms of orders of magnitude;
any greater precision is unjustified. For every important piece of
observed data, there are probably a dozen theories; in no field has
a theorist so much room to manœuvre. One new observation, such
as that of the 3° K microwave black-body radiation, can cause the
revision of many ideas, previously considered fundamental.

If after reading this short book, the reader retains the interest
in astrophysics that drew him to it, while at the same time he has
acquired sufficient background to follow the most exciting developments of the subject in the current literature, then the author will
have accomplished his objective in the writing of this work.

T. C. W.

1

Relevant Astronomical Vocabulary

1.1 Stellar studies

(A) THE SUN

Although the sun dominates the solar system, being both its parent and energy source, its principal importance in astrophysical studies is its proximity. It is a star of average size, is probably about 10^9 years old and will last at least another 10^{10} years. Its evolution has been quite normal with little variability; the only evidence of violent activity are the solar flares and sunspots. These phenomena, although violent, are surface disruptions and do not effect the star as a whole.

Because of its proximity the parameters associated with the sun can be accurately determined. Since these parameters are typical of a star, it is convenient to use them as astrophysical units. These quantities are summarized in Table 1.1, together with some convenient units of length.

TABLE 1.1

ASTRONOMICAL CONSTANTS

Mass of sun	$M_\odot = 1 \cdot 99 \times 10^{33}$ g
Luminosity of sun	$L_\odot = 3 \cdot 86 \times 10^{33}$ ergs/sec
Radius of sun	$R_\odot = 6 \cdot 96 \times 10^{10}$ cm
Astronomical unit (distance from earth to sun)	1 a.u. $= 1 \cdot 50 \times 10^{13}$ cm
Light-year	1 l.y. $= 9 \cdot 46 \times 10^{17}$ cm $\sim 10^{18}$ cm
Parsec	1 pc $= 3 \cdot 09 \times 10^{18}$ cm $\sim 3 \cdot 3$ l.y.

(B) STELLAR OBSERVATIONS

Stellar studies have been pursued almost exclusively through the medium of optical astronomy, most observations being confined by atmospheric absorption to wavelengths in the visible region (Fig. 1.1)

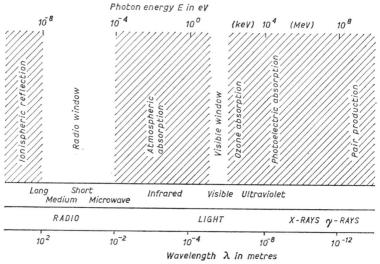

Fig. 1.1. Electromagnetic spectrum and terrestrial atmosphere

Considering the very narrow spectral range available, the range of data accumulated is remarkable. The data come from three sources: (1) the optical continuum, (2) line spectrum, (3) proper motions.

Because of the wide range of brightness observed in stars due both to the range of intrinsic brightness in the stars themselves and to the great range of distances, it is most convenient to measure brightness on a logarithmic scale. Since the first measurements were made using the human eye which is inherently logarithmic, this system was adopted at an early stage. The stars are measured in terms of magnitudes, each magnitude being approximately 2·5 times brighter than the next greater magnitude. The magnitude scale was originally chosen such that the brightest stars would have a magnitude of +1; magnitudes are now defined relative to a large number of secondary standards conveniently distributed over the celestial globe. On this scale the sun, because of its proximity, has a negative magnitude = −26·72. The brightest star, Sirius, has a magnitude of −1·50; the faintest stars visible with the naked eye have magnitudes of +6. The faintest stars detectable with the largest telescope in the world have magnitudes of +23. Note that the largest positive magnitudes correspond to the faintest stars.

The strict definition of magnitude is that a difference of five magnitudes corresponds to a difference of brightness of a hundred; one magnitude difference therefore corresponds to a factor of 2·512 difference in brightness. In practice magnitude must be defined in terms of the wavelength interval over which it is measured. The sensitivity of the human eye is best for green light; the modern method of measuring magnitude uses either photographic plates or photoelectric cells, with filters to define the waveband. It is convenient to define five magnitudes centred on wavelengths in the ultraviolet, blue, visual, red and infrared. Of these the most important are the ultraviolet U (3500 Å), the blue B (4500 Å) and the visual V (5500 Å). The colour index is the difference in the magnitudes of two colours.

To get the real, rather than the relative brightness, it is necessary to correct for the distance to the star. The absolute magnitude M is defined as the magnitude the star would have at a distance of 10 pc, i.e.,

$$M = m + 5 - 5 \log d, \qquad (1.1)$$

where d is the distance in parsec and m the apparent magnitude. The sun has an absolute magnitude of $+4 \cdot 86$; Sirius has an absolute magnitude of $+1 \cdot 36$.

The results of studies of the optical continuum of stellar systems can be conveniently summarized in the so-called Hertzsprung–Russell diagram. There are many forms of this diagram, the particular choice of

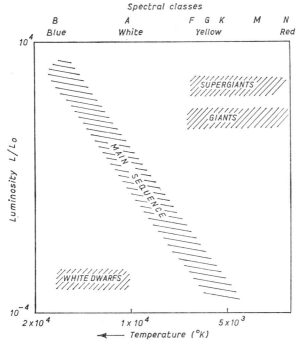

Fig. 1.2 *a*. Hertzsprung–Russell diagram

coordinates being a matter of convenience. Basically the ordinate represents the brightness or luminosity of the star, usually defined over a certain wavelength interval; the abscissa represents the temperature or colour of the surface (Fig. 1.2a). For each star a point is marked on the diagram, so that a map of stars is obtained in terms of brightness and surface temperature. Note that while brightness increases upwards, the temperature decreases from left to right (Fig. 1.2b).

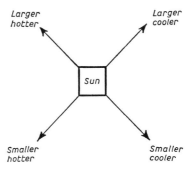

Fig. 1.2 *b*. Position of stars on H.R. diagram relative to the sun

An H.R. diagram may be drawn to represent the characteristics of all stars observed or to depict a select group. Since the former gives a hopelessly complex representation it is not very useful; this is not surprising since it contains stars of all ages, compositions and positions. A more useful representation is to impose some external criterion for the selection of a small group of stars. The variations within this group can then be studied with reference to their common characteristic. The

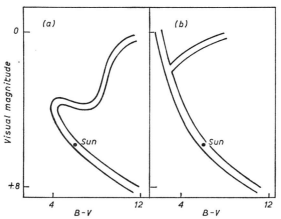

Fig. 1.3. H.R. diagram for (*a*) old cluster; (*b*) young cluster

easiest condition to impose is that only stars which are in a certain area in the sky be considered. Since many stars occur in clusters, this method restricts one to consideration of stars which probably had a common origin, i.e. all the stars condensed from a dense cloud of interstellar gas at about the same time. In clusters the ages and compositions of the stars will be approximately equal; because they are physically close to one another, their absolute magnitudes can be taken as proportional to their apparent magnitudes. The H.R. diagram for typical clusters is shown in Fig. 1.3: the coordinates used are convenient observational quantities.

The principal star types are named from their position on an H.R. diagram (Fig. 1.2a). The relative abundances of the principal stellar types are summarized in Table 1.2.

TABLE 1.2

Percentage occurrence of star types

TYPE	PERCENTAGE OCCURRENCE
Main Sequence	85
White dwarfs	10
Red giants	5
Variables	<0.1

The surface temperature of stars can be deduced from a study of the variation of intensity of the continuum with wavelength assuming the Planck radiation law. It is also possible to classify stars into spectral types; the usual system of classification is to consider ten types O, B, A, F, G, K, M, R, N, S. This sequence represents a decrease of temperature from O to S. There are many subdivisions.

From spectroscopic studies the relative abundances of the elements, the densities and the temperatures at the surface can be estimated. In all cases hydrogen and helium are the most abundant elements. The universal chemical abundances are discussed further in Chapter 4.

The abundance of hydrogen relative to helium is one of the most controversial aspects of this field at the moment; the sun, from which the most detailed spectroscopic information has been derived, is no help since its surface temperature is only 6000 °K, too cool to excite helium lines. Studies of hotter stars seem to indicate that the abundance of helium is less than would be expected, if currently popular cosmological concepts are to be believed. More recent observations point to a higher helium concentration.

From the study of proper motions, the Galactic stellar population can be broken up into two groups. Population I stars have low velocities

B

and their distribution in space corresponds to the distribution of the Galactic disc (Fig. 1.4*a*). Population II stars have a spherical distribution, with high velocities relative to the Galactic plane. They are believed to be old stars, with initial chemical compositions characteristic of the Galaxy at the time of its formation. Population I stars are relatively young which accounts for their distribution coinciding with the present shape of the Galaxy.

(c) STELLAR EVOLUTION

That branch of astrophysics that is concerned with the structure and evolution of stars is one of the most highly developed aspects of the discipline. While there are still some gaps in the picture, e.g. the birth and death of stars are still poorly understood, the overall situation is satisfactory, with theory keeping pace with observations. This happy state of affairs is all the more remarkable when one considers that the interior workings of the cores of stars are not directly observable; neutrino astronomy may provide the first direct observation of the sun's interior (Chapter 10).

The greatest single contribution to stellar studies was the elucidation by Bethe in 1939 of the mechanism by which the very large energies observed, are released. His proposal of thermonuclear reactions has since been elaborated and now forms the cornerstone of stellar models. The various cycles proposed have been studied in the laboratory and the cross-sections and rates of reaction are now well known. The impetus given to thermonuclear studies by Bethe's proposal is one of the outstanding contributions of astronomy to physics in this century.

The hydrogen burning reactions, which form the basis of the thermonuclear reactions, are well known; they are summarized in Chapter 10. Two cycles are of importance: the p–p cycle and the C.N.O. cycle. In the p–p cycle, four hydrogen atoms are fused at high temperatures to form helium:

$$4\,_1\text{H}^1 \rightarrow\ _2\text{He}^4 + 2e^- + \text{energy.} \qquad (1.2)$$

In the C.N.O. cycle, a helium atom is again formed from hydrogen atoms but the reaction proceeds via the formation of isotopes heavier than C^{12}, which must be present initially. Which of these two cycles is predominant depends on the temperature and hence on the mass of the star. In general in the more massive hot stars, the C.N.O. cycle is the most important. Under normal stellar conditions the burning of helium and other light elements is unimportant.

The first stage in star formation is the occurrence of a large irregularity in the distribution of the interstellar gas in the Galaxy. As this cloud condenses due to its gravitational forces, it splits into small units with masses of the order of stellar masses. Some of these clouds condense to form prototype stars, while the whole system constitutes a stellar cluster.

By considering the mechanical structure of stars and the conditions that govern equilibrium, it is possible to place upper and lower limits to the masses of gas that can form a star. In practice no stars have been observed with masses outside these limits. For equilibrium the forces due to thermonuclear energy generation in the star's core must balance the gravitational forces which tend to implode the star.

The total stellar energy, E, is given by

$$E = E_g + E_t, \qquad (1.3)$$

where E_g is the gravitational energy and E_t is the thermal energy. For a spherical object of total mass M, radius R and mean density

$$\rho = \frac{M}{(4\pi/3)R^3},$$

$$E_g = -\int_0^M \frac{Gm(r)\,dm}{r},$$

where $m(r)$ is the mass contained in the radius r and G the gravitational constant.

$$\sim -\frac{GM^2}{2R}$$

$$\sim -\frac{M}{2}G\rho(4/3\pi R^2). \qquad (1.4)$$

The thermal energy $E_t \sim 3MR'T$ where R' is the gas constant. Substituting in (1.3),

$$E = M[3R'T - (2\pi/3)GR^2\rho]. \qquad (1.5)$$

For a star to be bound, E must be <0. Hence if T is large, the density must also be large.

Thermonuclear reactions will only occur when $T \gtrsim 10^7\,°K$. To increase the temperature to the point at which nuclear reactions can take place, a minimum mass is required. This turns out to be $M_{min} > 0.02\,M_\odot$. At very large masses the radiation pressure must also be taken into account. When this exceeds the gas pressure, the star will be unstable and will oscillate with the ejection of mass until a stable configuration is achieved. The upper mass limit is believed to be $M_{max} \lesssim 30\,M_\odot$. Recently, the existence of condensed objects (superstars) with masses of 10^5 to $10^8\,M_\odot$ have been proposed. Preliminary estimates of the conditions under which they can exist have been made; there is some hope that they may be intimately related to the nucleus at the centre of the galaxy. Since these massive stars more properly belong in the realm of galaxies, they will be considered in this context in Chapter 7.

When a star is formed it will contract due to gravitational forces. At this stage heat transfer in the star is almost wholly convective. It will move towards the bottom left quadrant of the H.R. diagram, until it reaches the Main Sequence. Most of its lifetime is spent here, with hydrogen burning taking place in its core. The time spent on the Main Sequence depends on the mass and is roughly $10^{10} (M_\odot/M)^3$ years. To a first approximation then, the lifetime of a $1 \cdot 0 \ M_\odot$ star is 10^{10} years, that of a $10 \ M_\odot$ star is 10^7 years. As the hydrogen in the core becomes used up, the central temperature rises and the outer shell begins to expand. Hence the star evolves towards the right of the H.R. diagram since its surface temperature falls. However, the central temperature continues to rise and eventually helium burning takes place. Subsequent evolution depends on the mass; the heavier stars evolve to the red-giant phase where the lighter elements are burned and the outer envelope of the star is ejected explosively.

The lighter stars pass back through the Main Sequence where they eventually cool. Stars eventually reach the white-dwarf state, which is a stable, but cold, configuration. This is a highly collapsed stage, the radius being typically that of the earth, i.e., $\sim 10^{-3} \ R_\odot$; the luminosity is low so that the white dwarfs are barely visible and difficult to study.

At the high densities that exist in a white dwarf, most of the matter is degenerate, i.e. the energy is as low as the Exclusion Principle will allow and particles are confined to volumes of the order of their Compton wavelengths. Since no two fermions may occupy the same energy state unless they have opposite spins, a Fermi pressure will exist, even at zero temperatures, and balance the gravitational forces. At these densities the only fermions of importance are electrons; the neutron and proton only become important at higher densities because of their smaller Compton wavelengths. The existence of the ultimate collapsed state, the neutron star, where densities of 10^{13} g/cm^3 prevail and the star behaves like a giant nucleus, has been proposed but these objects have not yet been observed (see Chapter 8).

White dwarfs have, in fact, a finite temperature (their surface temperature is greater than that of the sun) but to a first approximation they can be treated as cold stars, with all the pressure coming from the Fermi pressure. An analysis of such stars shows that they exist only if

$$M < 5 \cdot 6 (Z/A)^2 \ M_\odot, \tag{1.6}$$

where Z is the atomic number and A the atomic weight of the material in the star. For a helium star, $Z = 2$, $A = 4$, so that M must be less than $1.4\ M_\odot$. Heavier stars than this must lose mass before they can evolve to this cold state.

1.2 Galaxies

The sun is one of 10^{11} stars in the complex system known as the Galaxy. Our presence in the Galaxy complicates its study although it is possible to deduce much of the general structure by analogy with other galactic systems. Given the broad outlines of the Galaxy, in particular the off-centre position of the sun, the details can be filled in by direct observation. The presence of interstellar dust, particularly near the centre of the Galaxy, has limited the application of optical astronomy. The advent of radio astronomy and particularly the discovery of the 21 cm line from neutral hydrogen has led to the present picture of the Galaxy. This is fairly satisfactory although there are still great uncertainties; the Galactic centre is still virtually unexplored.

The general features of the Galaxy, as seen in cross-section, are shown in Fig. 1.4a. Most of the stars are contained in the Galactic disc

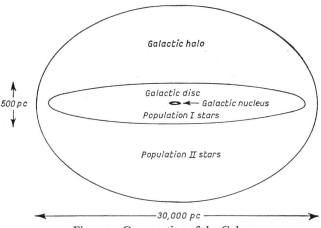

Fig. 1.4a. Cross-section of the Galaxy

which has a thickness of a few hundred parsec. These are mainly Population I stars, which have been formed comparatively recently. Population II stars have a spherical distribution similar to the distribution of gas in the early stages of galactic formation. This cloud of gas gradually flattened to the observed disc. The density of stars in the disc is about $0.1\ (pc)^{-3}$.

The disc itself, seen from above, is depicted in Fig. 1.4b. The spiral

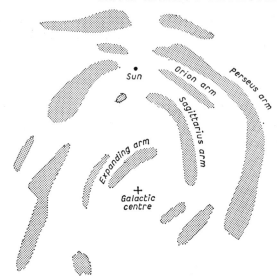

Fig. 1.4*b*. Spiral structure of Galaxy as outlined by hydrogen concentration

structure is immediately apparent. The sun is situated on the Orion arm, about 10 kpc from the Galactic centre. The structure of the arms is somewhat difficult to analyse because of the sun's position; the arms can however be traced from the distribution of the bright O and B stars and from radio observations of the 21 cm line from the H I regions (neutral hydrogen) which occupy the arms. Ionized hydrogen (H II) is also observed in the Galactic disc as are a number of other molecules. The density of gas in the disc is about 10^{-24} g/cm³. From the Doppler shift of the 21 cm line the radial velocities of the arms may be deduced; these velocities suggest that the arms would have their present properties, if they were radial but straight 4×10^8 years ago.

The Galactic core or nucleus has a radius of about 10 pc. The stellar density is believed to be high in this region (10^2 (pc)$^{-3}$) but direct optical observations are impossible because of the interstellar absorption of light. External galaxies are observed to have nuclei; in Andromeda the nucleus contains 10^7 solar masses. The radio source, Sagittarius A, is coincident with the Galactic centre and recently a group of x-ray sources have been found to cluster in this general area.

The Galactic halo is one of the more uncertain features of the Galaxy. Although some external galaxies appear to have halos, the evidence for the Galactic halo is somewhat scanty. The density of the halo would be about 10^{-26} g/cm³, consisting mostly of ionized hydrogen. It is roughly spherical and is believed to contain a weak magnetic field. This field has two effects: (*a*) it causes most of the cosmic radiation emitted in the

Galaxy to be trapped, (*b*) relativistic electrons trapped by this field radiate by the magnetic bremsstrahlung mechanism. This radiation constitutes about 75% of the radio cosmic background. The existence of the Galactic halo is critical for theories of the origin of the cosmic radiation (see Chapter 4). Some of the parameters associated with the Galaxy are summarized in Table 1.3.

TABLE 1.3

Galactic parameters

Mass of Galaxy	$= 10^{11} M_\odot = 2 \times 10^{44}$ g
Age of Galaxy	$= 10^{10}$ years
Optical luminosity	$\sim 4 \times 10^{43}$ ergs/sec
Magnetic field strength	$\sim 3 \times 10^{-6}$ gauss

The Galaxy belongs to the Local Cluster, a group of 15–20 galaxies contained in a radius of 0·7 Mpc. The biggest galaxy in the cluster is the Andromeda Nebula (about $4 \times 10^{11} M_\odot$) which is only 300 kpc away from the Galaxy. The nearest objects are the large and small Magellanic Clouds; these are about one tenth the size of the Galaxy and are only 50 kpc away. It is not clear whether they are so close by chance or whether they are intimately associated with the Galaxy. The Local Cluster is believed to belong to a large group of 200 galaxies known as the Super Cluster.

External galaxies recede with velocities which increase with distance *d*, according to Hubble's Law

$$v = Hd, \qquad (1.7)$$

where $H =$ Hubble's constant ~ 100 km/sec/Mpc $\sim 3 \times 10^{18}$/sec and $d =$ distance from the Galaxy. When $v = c$, the radiation can no longer reach the Galaxy; this distance is the edge of the observable universe and is equal to 10^{28} cm. The age of the universe is usually taken as the reciprocal of H, i.e., $\approx 10^{10}$ years.

The density of galaxies does not appear to differ appreciably with increasing *d*. The most distant galaxies can most easily be studied with radio techniques due to the comparative insensitivity of optical telescopes. The general classification of galaxies by their optical properties is given in Table 1.4. The radio properties will be considered in Chapter 5. The Galaxy and the Andromeda Nebula are classified as Sb and are exceptionally massive.

Unlike its counterparts on a stellar scale, the study of galactic formation and evolution is far from complete. It is still a mystery why galaxies should have the sizes observed and what the conditions governing the condensation of the tenuous extra-galactic gas to the point of

TABLE 1.4

Classification of galaxies

	SUB-CLASSIFICATION	OCCURRENCE (%)	MASS
Elliptical E_n	n where $n/10$ = ellipticity	30	$10^{12} M_\odot$
Spiral S	⎰ a, b, c, in order of decreasing	30	10^{10} to $10^{11} M_\odot$
Barred Spiral SB	⎱ ratio of core to arms	30	10^{10} to $10^{11} M_\odot$
Irregular I_r		10	

galaxy formation are. It is thought that the gas initially condenses as a spherical system. Because of rotation this gas cloud will flatten to an ellipsoid and eventually to a disc. Once stars are formed, this flattening will tend to slow down and the Galaxy will retain the optical appearance that it had at the time of star formation. The critical factor is thus the point at which star formation takes place. The conditions governing this point are unknown. The role of the galactic nucleus is also unknown. In some galaxies violent explosions are observed in the nucleus; these may be exceptional galaxies or this may be a normal stage in galactic evolution. This problem is considered again in Chapters 5, 6 and 7.

2
Magnetic Bremsstrahlung and Compton Radiation

Among the many electromagnetic phenomena observed in astrophysical situations, two processes are of particular interest. The magnetic bremsstrahlung mechanism has proved to be of fundamental importance in radio astronomy, where it appears to be the dominant mechanism; the Compton mechanism may prove to be its counterpart in x-ray and gamma-ray astronomy.

2.1 Magnetic bremsstrahlung

At an early stage in the development of radio astronomy it became evident that much of the radiation observed was non-thermal in character. In 1953 it was proposed that the same mechanism that is observed in large synchrotron accelerators to limit the acceleration of electrons, might be applicable to astronomical objects. This 'synchrotron' radiation, or magnetic bremsstrahlung, as it is more properly called, occurs when relativistic particles traverse magnetic fields. In laboratory accelerators electrons with energies of about 1 MeV are observed to radiate at radio frequencies of about 10^3 Mc/s in magnetic fields of 10^2 gauss. The strength of the magnetic fields in astronomical objects is still uncertain but present estimates are in the range 10^{-3} to 10^{-5} gauss. Electrons with energies of about 10^2 GeV in fields of 10^{-4} gauss will also radiate at the above frequency. Hence, although the astronomical situation involves weak magnetic fields, the presence of electrons of relativistic energies is sufficient to give detectable radio emission. Relativistic electrons are also detected in the cosmic radiation (see Chapter 4).

A non-relativistic electron moving through a region of uniform magnetic field will move in a spiral whose axis is parallel to the magnetic field lines (Fig. 2.1a). The angular frequency is given by:

$$\nu_L = \frac{1}{2\pi} \frac{eH}{mc},$$ (2.1)

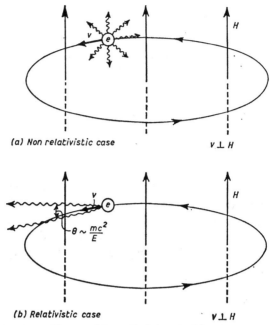

(a) Non relativistic case $v \perp H$

(b) Relativistic case $v \perp H$

Fig. 2.1. Magnetic bremsstrahlung

where e = electronic charge, H = magnetic field, m = electron mass and c = velocity of light. The electron then radiates like a dipole with frequency ν_L.

If the energy of the electron, E, is relativistic, then the radiation will be no longer isotropic but will be concentrated in a cone of angle

$$\theta = \frac{mc^2}{E} = \frac{1}{\gamma},$$

where γ = Lorentz factor = E/mc^2. The axis of the cone will be centred on the electron trajectory at that instant; as the electron orbits, it will radiate in all directions within $\theta/2$ of its orbital plane (Fig. 2.1b). An observer situated in this plane will receive radiation only as long as he lies in this cone, i.e., he will observe pulses of duration

$$\mathrm{d}t = \frac{R\theta}{c}\frac{1}{\gamma^2}, \tag{2.2}$$

where R is the radius of curvature of the electron trajectory $\sim \dfrac{c}{2\pi\nu_L}$. Also in the relativistic case,

$$\nu_L = \frac{1}{2\pi}\frac{eH}{mc}\frac{1}{\gamma} \quad \text{and} \quad \mathrm{d}t = \frac{mc}{eH}\frac{1}{\gamma^2}. \tag{2.3}$$

The frequency of occurrence of these radiation pulses as seen by the

observer is ν_L, where the motion of the electrons parallel to the field
has been ignored. The radiation will have a continuous distribution,
the harmonics of the fundamental frequency, ν_L. All harmonics will not
be equally present; the maximum will occur at $\nu_m \sim 1/dt$. The effect
of the relativistic energy has been to change the spectral distribution
of the radiation from a line at ν_L to a continuous distribution. The form

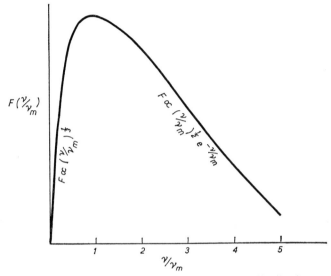

Fig. 2.2. Magnetic bremsstrahlung frequency distribution

of the spectral distribution $F(\nu/\nu_m)$ is shown in Fig. 2.2. It is asym-
metric but can be represented by the following asymptotic expressions
[Morrison (1961)];

$$F\left(\frac{\nu}{\nu_m}\right) = 2\cdot15\left(\frac{\nu}{\nu_m}\right)^{1/3}, \qquad \nu \ll \nu_m$$

$$F\left(\frac{\nu}{\nu_m}\right) = 1\cdot26\left(\frac{\nu}{\nu_m}\right)^{1/2} e^{-\nu/\nu_m}, \qquad \nu \gg \nu_m. \qquad (2.4)$$

In practice for most astrophysical purposes it is sufficient to consider all
the radiation from an electron of Lorentz factor $\gamma = E/mc^2$ as having
a frequency

$$\nu_s = \frac{3}{2}\nu_m = \frac{3}{2}\frac{1}{2\pi}\cdot\frac{eH}{mc}\gamma^2$$
$$\approx 4\cdot2.10^{-6}\gamma^2 H, \qquad (2.5)$$

where ν_s is in Mc/s, H is in microgauss [Felten and Morrison (1966)].
This radiation causes an energy drain on the electrons so that E

steadily decreases. In some situations this energy drain is the most serious energy loss mechanism. The power radiated per second by an electron of Lorentz factor γ in a magnetic field with a value H perpendicular to the electron trajectory is given by:

$$P(\gamma,H) = \frac{2}{3}r_0{}^2c\gamma^2H^2, \qquad (2.6)$$

where r_0 is the classical electron radius. Using the assumption that all the power is radiated at a frequency ν_s (equation 2.5), the magnetic bremsstrahlung frequency spectrum is given by:

$$P(\nu,\gamma,H) = P(\gamma,H)\delta(\nu - \nu_s). \qquad (2.7)$$

This is, of course, monochromatic. In practice the angular distribution of the radiation and the parallel component of the magnetic field must also be taken into account. If the region considered is sufficiently large that the fields and particles are isotropically distributed, then the radiation can also be considered isotropic.

In astronomical sources radiation from a whole spectrum of electron energies must be considered. The electrons in general, will have a power law distribution:

$$n(\gamma)\,d\gamma = n_0\gamma^{-m}\,d\gamma.$$

In a region of linear dimension L over which the magnetic field is uniform, but randomly oriented, the intensity received at earth in watts/m²/ster/c/s is given by:

$$I(\nu) = 4{\cdot}8 \times 10^{-20}(4{\cdot}9 \times 10^2)^{3-m}n_0LH^{(1+m)/2}\nu^{(1-m)/2}, \qquad (2.8)$$

where n_0 is in c.g.s. units, L is in light-years, H is in μ gauss and ν is in Mc/s [Felten and Morrison (1966)]. The most important feature of the expression is that the frequency spectrum is a power law, i.e.,

$$I(\nu) \propto \nu^{-\alpha},$$

where $-\alpha = (1 - m)/2$ or $m = 2\alpha + 1$. α is called the spectral index. By making observations at a number of frequencies, α and hence m, the exponent of the electron distribution, can be determined.

One of the most characteristic features of the magnetic bremsstrahlung process is that the radiation is polarized. In general the electric vector of the radiation from an accelerated electron will have its maximum value in the direction of acceleration; this direction will depend on the direction of the magnetic field. When a system of electrons is considered, the degree of polarization depends on the degree of uniformity of the magnetic field. No polarization will be observed if the fields are completely random.

Polarization has been observed in a number of sources at both radio and optical frequencies. This is regarded as strong evidence in favour of the magnetic bremsstrahlung radiation mechanism as the major source of non-thermal radio emission. It also indicates some degree of uniformity in the distribution of astronomical magnetic fields.

2.2 Relativistic particle energies

If the radiation from a discrete radio source is attributed to the magnetic bremsstrahlung mechanism, then the total energy in the source in the form of relativistic electrons can be estimated. It is necessary to assume that the magnetic field and the electron density are uniform over the source. Equation (2.8) can be written

$$I(\nu) = k(m)\frac{n_0 V_0}{d^2}H^{(1+m)/2}\nu^{(1-m)/2}, \tag{2.9}$$

where $k(m)$ contains some constants and is a known function of m, and

$V_0 =$ total volume of the radio source, $= \frac{4}{3}\pi(R/2)^3$;

$d =$ distance to the source, $= R/\theta$ where $\theta =$ angular diameter of the source.

The total number of relativistic electrons is given by

$$\mathcal{N} = \int_{\gamma_1}^{\gamma_2} n_0 V_0 \gamma^{-m} \, d\gamma,$$

where γ_1 and γ_2 are determined from the limits of the radio spectrum. Similarly the total energy in the source in the form of relativistic electrons is given by:

$$U_e = \int_{\gamma_1}^{\gamma_2} n_0 V_0 \gamma^{-m}\gamma \, d\gamma . m_0 c^2.$$

This can be shown to reduce to:

$$U_e = f(m,\nu)R^2 I(\nu)H^{-3/2}, \tag{2.10}$$

where $f(m,\nu)$ is a known function of m and ν. $I(\nu)$ and m can be determined directly from radio observations; R can be estimated using redshift measurements for extra-galactic sources. If H is known, then U_e the total energy in relativistic electrons in the source, can be deduced. Unfortunately there is, at present, no direct method of measuring H; an indirect method based on equipartition arguments can be used to get a rough estimate of H.

Let U_p be the total energy in the source in the form of relativistic particles; these, in general, will be either protons or electrons. Although radio emission is only expected from electrons (since $P(\gamma,H) \propto \gamma^2$) the presence of protons can be inferred from the cosmic radiation. If the electrons and protons are genetically related, then

$$U_p = kU_e,$$

where k is a numerical factor. In the cosmic radiation k is almost 100; usually values between 1 and 100 are used for k for energy estimates if no other information is available.

$$U_p = kf(m,\nu)R^2I(\nu)H^{-3/2}$$
$$= K_1 H^{-3/2}.$$

Let U_H be the magnetic energy density in the source which is assumed uniform over the entire volume V_0.

$$U_H = \frac{H^2}{8\pi}V_0 = K_2 H^2.$$

The total energy in the source is given by:

$$U = U_p + U_H = K_1 H^{-3/2} + K_2 H^2.$$

The total energy is a minimum when $dU/dH = 0$, i.e.,

$$-\frac{3}{2}K_1 H^{-5/2} + 2K_2 H = 0,$$

or
$$U_p = \frac{4}{3}U_H.$$

Hence the minimum total energy occurs when the magnetic energy density and the relativistic particle energy density are of the same order. If $U_H < U_p$, then it would not be possible to contain the particles in the source. In most astrophysical calculations equipartition of energy is assumed between U_H and U_p.

Therefore
$$kf(m,\nu)R^2I(\nu)H^{-3/2} = \frac{4}{3}\frac{H^2}{8\pi}V_0,$$

or
$$H = \left(6\pi kf(m,\nu)R^2\frac{I(\nu)}{V_0}\right)^{2/7}. \tag{2.11}$$

This rough estimate of H is limited by the necessity of choosing arbitrarily a value for k and assuming equipartition.

Some applications of this method of calculation will be considered in later chapters.

2.3 Compton scattering

In Compton scattering a photon collides with an electron which can be assumed to be at rest; the photon trajectory is changed, its energy degraded and the electron takes up the energy difference (Fig. 2.3a). The inverse process, in which the photon energy is much less than that of the electron, results in the transfer of energy from the electron to the photon; it can thus be considered a process of photon enhancement. Under laboratory conditions this second process is unimportant as photon densities are usually low and electron energies small. Since

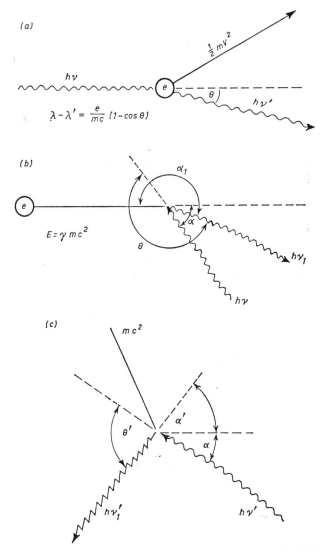

Fig. 2.3. (a) Compton scattering; (b) Compton scattering in the LAB frame; (c) Compton scattering in the rest frame of the electron

electron densities are high in high-density materials, the Compton effect is an important scattering process for medium-energy gamma-rays.

In the astrophysical situation conditions are reversed; photon densities are relatively high, electron densities low and the energy spectrum of cosmic electrons known to obey a power law up to energies of 10^{11}

to 10^{12} eV. The importance of this process in high-energy processes in astrophysics has recently been recognized; it can be considered from the viewpoint of (a) an electron energy degradation process and (b) an x-ray and gamma-ray production process.

The physical process has been treated in detail by Donahue (1951). Felten and Morrison (1966) have recently reviewed the relevance of the process for astrophysics. In the laboratory frame of reference S, i.e., the coordinate system with reference to the fixed stars, the collision can be represented as in Fig. 2.3b. The notation is as in section 2.1; in addition $\beta = v/c$ where v = electron velocity; α = angle which incident photon makes with electron trajectory; α_1 = angle which recoil photon makes with electron trajectory; $h\nu$ = incident photon energy; $h\nu_1$ = recoil photon energy; $\theta = \alpha + \alpha_1$ = scattering angle of photon. For $\beta \sim 1$, $h\nu_1 > h\nu$. If the collision is viewed in the rest frame of the electron S', then it reduces to a simple Compton collision (Fig. 2.3c) where quantities in S' are now shown as dashed. The usual relativistic transfer-equations can now be employed:

$$h\nu' = \gamma h\nu (1 + \beta \cos \alpha),$$
$$h\nu_1 = \gamma h\nu_1'(1 - \beta \cos \alpha),$$
$$\tan \alpha' = \frac{\sin \alpha}{\gamma(\cos \alpha + \beta)}.$$

Using the Compton equation in the S' frame,

$$h\nu_1' = \frac{h\nu'}{1 + (h\nu'/mc^2)(1 - \cos \theta')},$$
$$h\nu_1 = \frac{\gamma^2 h\nu(1 + \beta \cos \alpha)(1 - \beta \cos \alpha_1')}{1 + (\gamma h\nu/mc^2)(1 + \beta \cos \alpha)(1 - \cos \theta')}. \qquad (2.12)$$

For the extreme relativistic case, β tends towards 1, $\gamma \gg 1$, so that α' goes to zero and α to π. Physically this corresponds to the recoil photons taking the direction of the incident electrons. For these conditions all the terms in brackets in equation (2.12) can be neglected.
Therefore

$$h\nu_1 \approx \gamma^2 h\nu, \qquad (2.13)$$

where $\gamma \gg 1$ but $\gamma . h\nu \ll mc^2$. This scattering can be treated as classical Thomson scattering of radiation by electrons. The cross-section for the process is the Thomson cross-section:

$$\sigma_T = \frac{8\pi}{3} r_0^2, \qquad (2.14)$$

where r_0 = classical electron radius = e^2/mc^2. The number of collisions per second, N, that a relativistic electron undergoes in traversing an isotropic photon distribution with energy density ρ is given by,

$$N = \sigma_T c \frac{\rho}{h\nu}. \qquad (2.15)$$

Hence the power lost per second by an electron with Lorentz factor γ is given by:

$$P(\gamma,\rho) \approx \gamma^2 h\nu\sigma_T \frac{c\rho}{h\nu} \approx \gamma^2\sigma_T c\rho.$$

A more rigorous treatment gives a numerical factor of $4/3$ so that

$$P(\gamma,\rho) = 2\cdot7 \times 10^{-14}\gamma^2\rho. \tag{2.16}$$

To calculate the resulting photon spectrum the spectrum of energies from a single scattering should be calculated. In practice for all cases of interest, a power law energy spectrum of electrons is involved; viz.,

$$n(\gamma).d\gamma = n_0\gamma^{-m}\,d\gamma.$$

The recoil photon spectrum can then be represented by:

$$P(\nu,\gamma,\rho) \approx P(\gamma,\rho)\delta(\nu - \nu_c), \tag{2.17}$$

where ν_c is defined by: $h\nu_c \sim \dfrac{4}{3}\gamma^2.h\nu.$

This approximation is similar to that used in the treatment of magnetic bremsstrahlung (eqn 2.5) where the radiation is assumed monochromatic.

A further approximation is to consider the incident photon spectrum as represented by an average photon energy; for a black-body spectrum this energy is given by

$$\overline{h\nu} = 2\cdot7kT,$$

where $k =$ Boltzmann's constant, $T =$ black-body temperature.

If the electron and photon distribution can be considered as homogeneous and isotropic over the region for which the interaction is considered, then a useful expression can be used for the specific intensity [Felten and Morrison (1966)]:

$$I(h\nu) \sim 10^3(56\cdot9)^{3-m}n_0L\rho\,T^{(m-3)/2}(h\nu)^{(1-m)/2}$$
$$\text{eV}/\text{eV cm}^2 \text{ sec ster}, \tag{2.18}$$

where $L =$ distance in light-years along the line of sight through the region of interaction, n_0 is in c.g.s. units, ρ is in eV/cm^3, T in degrees Kelvin, and $h\nu$ in eV.

The condition $\gamma h\nu < mc^2$ restricts the application of this simple treatment. For $h\nu \sim 1$ eV (optical) and $E \sim 10^{10}$ eV ($\gamma \sim 2\cdot10^4$), $\gamma^2.h\nu \sim 4\cdot10^8$ eV. The above treatment can thus be used over much of the x-ray and lower energy gamma-ray spectrum. For $\gamma h\nu > mc^2$ the Klein–Nishina high-energy approximation can be used in place of the Thomson cross-section:

$$\sigma_{\text{K.N.}} = \pi r_0^2 \frac{mc^2}{\gamma.h\nu} \ln\left(\frac{2\gamma h\nu}{mc^2}\right). \tag{2.19}$$

c

The mean energy loss per scattering is now independent of the incident photon energy and is comparable with the electron energy:

$$\overline{h\nu_1} \sim \gamma . mc^2. \tag{2.20}$$

Hence the rate of electron energy loss and the recoil photon energy spectrum can be derived as before.

3
Novae and Supernovae

3.1 Variable stars

Although stellar phenomena are distinguished by the degree to which they conform to the evolutionary pattern, which is summarized in the H.R. diagram, the relatively small number of stars that are exceptions to this steady evolution are particularly interesting both for the phenomena that they exhibit and for the insight that they give into the conditions necessary for stellar equilibrium. The most spectacular of these stars are novae and supernovae; since they are comparatively rare and short-lived, they are difficult to study. Other classes of non-steady stars are more common and because the variations that they exhibit are either almost static or repetitive, a detailed study is possible. The most important of these are:

(A) PULSATING VARIABLES

All stars are variable to some degree. The pulsating variables are those whose light curves (luminosity versus time, Fig. 3.1) show a regular

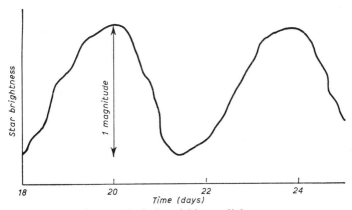

Fig. 3.1. Typical variable star light curve

change in amplitude with almost constant period. Historically the discovery that there was in these variables a definite relationship between luminosity and period (the longer the period, the greater the luminosity) led to the establishment of a distance scale to stellar clusters within the Galaxy. The stars that exhibit this regular variability are usually giants or supergiants; they can be either Population I or II stars and spectral classes from O to M have been observed. The periods range from a few hours to a year and the amplitude of the luminosity change can be anything between 0·1 and 6·0 magnitudes. The pulsations are probably oscillations about some equilibrium position of mechanical equilibrium.

(b) T TAURI STARS

A number of stellar types display irregular luminosity changes; the most important of these are the T Tauri stars. Each of these is surrounded by a thin nebula which suggests the ejection of mass at some stage. The continuum emission is particularly strong in the ultraviolet and may be non-thermal; one explanation is that it is due to magnetic bremsstrahlung radiation although no radio emission has been detected. The T Tauri stars are believed to be young stars of intermediate mass in the process of formation from dense interstellar gas.

(c) MAGNETIC VARIABLES

The Zeeman splitting of spectral lines in stars allows the magnetic field to be deduced. In one type of star, very strong fields (up to 30,000 gauss) are observed over large regions; in some cases these fields vary rapidly and irregularly with periods of about a week. Zeeman splitting can only be measured where the spectral lines are sharp; there may be stars with even larger fields which have not yet been detected. The mechanism by which these strong fields are generated and varied in such short periods is not yet understood.

(d) FLARE STARS

Even the sun, which is normally quiescent, displays violent local disturbances. Flares which are similar in character to the solar disturbances but on a larger scale have recently been observed from a special class of stars called Flare stars. Simultaneous surveys of likely stars have been made at optical and radio wavelengths. The results show that the flares are similar to Type I solar flares. The optical emission is 10 to

100 times greater than a solar flare; the radio emission is 10^4 to 10^6 times greater. The maximum increase in luminosity observed is six magnitudes. These are recurrent phenomena, but seem to occur only in red dwarfs, which are among the most common stars in the Galaxy. It is probable that their origin is similar to that of solar flares involving electromagnetic disturbances in the outer layers of the star.

3.2 Novae

Stars which increase their optical luminosities by many magnitudes over a short period of time can be easily identified by the naked eye. These dramatic changes were carefully recorded by early astronomers because of their supposed importance in terrestrial affairs. 'Guest stars', as novae were called (to distinguish them from 'sweeping stars', i.e. comets), are frequently recorded in the Chinese chronicles and the accuracy of these recordings is a tribute to methodical observers working without the aid of instruments. Hence the study of novae is as old as astronomy.

The principal feature of novae is a rapid increase in optical brightness over a few days followed by a slow decline over a period of months. The initial rapid change corresponds to an explosion in the star with the rapid ejection of a considerable fraction of the star's mass. The appearance of the star after the outburst has spent itself is not significantly different from its appearance prior to the explosion.

As yet all the observational data about novae have come from optical astronomy. No radio emission has been detected; there is some evidence for the association of 'x-ray stars' with old novae (see Chapter 8). Despite the limited region of the electromagnetic spectrum that can be used for their study, the large numbers of novae that have been observed have enabled an impressive array of observational data to be accumulated. The principal data that any theory of novae origins must explain is listed below.

(*i*) *Energy*. The mass ejected as a result of the outburst can be as much as 10^{28} g. This is ejected with velocities of 1000 km/sec, corresponding to a total kinetic energy of 10^{44} ergs. The total energy radiated is 10^{44} to 10^{45} ergs. The overall change in visual magnitude can be as much as eleven.

(*ii*) *Light curve*. Although the light curves of individual novae differ in their finer details, several distinct features can be recognized. An initial rise through nine magnitudes is followed by a temporary halt and then a final slower increase of two magnitudes to the maximum (Fig. 3.2). Once the maximum is reached, the light curve falls off fairly

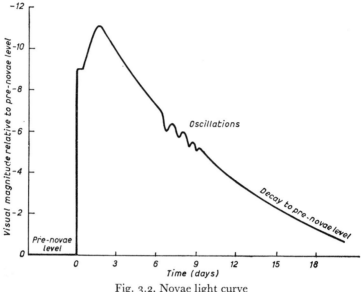

Fig. 3.2. Novae light curve

rapidly. Three stages can be distinguished in the decline: a steep decline of 3–4 magnitudes, a transition period during which oscillations may take place, a final slow decline to the post-novae star. Generally the brightest novae decay the fastest. The novae that decline slowly tend to be more varied in all stages of their development.

(*iii*) *Stellar type*. In the H.R. diagram the pre-novae stars lie in the region below the Main Sequence and above the white dwarfs. They are fairly faint blue stars probably with $M \sim M_\odot$ and $T \sim 5 \times 10^4$ °K. The post-novae are similar and the phenomenon may be recurrent with long intervals between outbursts. From the spatial distribution novae appear to be associated with Population II stars.

(*iv*) *Spectral features*. Absorption and emission lines are seen against a variable continuum. Although detailed interpretation is not possible, the general features can be recognized. After the maximum, bright emission lines are apparent; these lines persist after the continuum has decreased significantly. They come primarily from the gas shell that surrounds the star.

(*v*) *Occurrence*. The rate at which novae occur in the Galaxy is an important parameter which is still in dispute. Estimates vary from 50 to 200/year/galaxy.

The early theories of the novae mechanism were that the phenomenon represented the results of a stellar collision or the passage of a star through a gaseous nebula. These theories proved inadequate and were replaced by theories which associated novae with rotational instabilities in binary systems. This instability arises if the two components of the binary separate; to attain stability one of the components must eject matter. A more detailed analysis has shown that although this mechanism may be important for other variables, it is not the dominant process in novae.

Schatzman (1965) has developed a theory in which the novae explosion is due to the nuclear processes in the star. In the course of stellar evolution there are a number of different phases of fusion which could give rise to non-equilibrium conditions. Several reactions have been proposed as possible detonators of the explosion; from the position of the pre-novae on the H.R. diagram the burning of light elements should be involved. The sudden release of energy at the centre of the star, due to the onset of helium burning, could release enough energy to cause a shock wave to propagate outwards carrying with it a large amount of stellar material. With this model agreement is found between theory and observation for such parameters as the period of recurrent novae and the total energy release/outburst.

3.3 Supernovae

Although the study of novae dates back to ancient times, it was not until comparatively recently that it was recognized that these dramatic events are transcended by a less frequent, but more drastic, stellar explosion. Supernovae are brighter than novae by many magnitudes but could not be distinguished from the more frequent novae until distance measurements, and hence absolute magnitudes, were available. A plot of frequency of occurrence as a function of apparent novae magnitude at maximum gives a smooth distribution (Fig. 3.3a) with no indication of two distinct classes of object. With the development of more powerful telescopes and the identification of stellar explosions in extra-galactic systems, it was possible to plot the frequency of occurrence as a function of absolute magnitude (Fig. 3.3b). The distinct class of very bright novae were called supernovae. The last recorded observation of an event in this category in the Galaxy was in A.D. 1604; it is not surprising therefore that the phenomena should have escaped detection until optical astronomy was extended to the study of systems outside the Galaxy.

Over a hundred supernovae have now been detected, due largely to regular surveys at a number of observatories in the last thirty years. These optical observations of the growth and first stages of the decline of supernovae in external galaxies, taken with the observations on

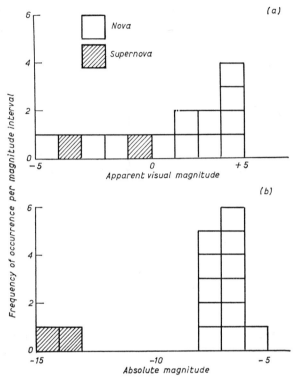

Fig. 3.3. Separation of novae and supernovae into two classes by frequency magnitude diagram

known supernovae remnants in the Galaxy in the radio, optical and x-rays bands, amount to an impressive array of data against which models of the catastrophic stellar explosion can be compared. One of the brightest supernovae in an external galaxy was detected by Zwicky in 1937 in the dwarf galaxy IC 4182. At maximum this event had an apparent magnitude of +8·2; the distance to the galaxy is 3 million light-years so that the absolute magnitude of the supernova was −16·6. This is over a hundred times brighter than its own galaxy which contains a million stars. One year after the explosion the luminosity had decreased by a factor of 10^6.

Three supernovae have been observed in the Galaxy. Of these the Crab Nebula, which was recorded by Chinese astronomers in A.D. 1054 as a 'guest star' of unusual brightness, was the closest (1100 parsec). The other supernovae, that of Tycho Brahe in A.D. 1572 and of Kepler in A.D. 1604, were named after the famous astronomers who observed them. One of the early surprising results of radio astronomy was that the Crab Nebula, now not visible to the naked eye at optical wavelengths, is coincident with one of the strongest radio sources, Taurus A. Radio emission has also been detected from the remnants of SN 1572 and 1604. Cassiopeia A, the brightest radio source in the sky, was not observed optically as a supernovae; this Galactic radio source has all the radio characteristics of one and it is believed that the optical event may have been obscured by interstellar dust. Recently the Crab Nebula has been positively identified with one of the bright x-ray sources; TAU XR-1.

At the maximum the supernovae luminosity is at least two magnitudes greater than the luminosity of the brightest nova (about -8 absolute magnitude). Supernovae can have absolute magnitudes as high as -19.

Supernovae are not peculiar to any one type of galaxy. There is some evidence for a higher rate of occurrence among Sb and Sc spiral galaxies and irregular galaxies. This interpretation is confused by the tendency of observed supernovae to lie on the outside edges of the galaxy. In dense regions such as elliptical galaxies or the nuclei of spiral galaxies, interstellar dust may obscure the optical evidence for the explosion.

One of the most important supernovae parameters is their rate of occurrence per galaxy. From the observation of supernovae in external galaxies the rate is estimated as $1/300-400$ years/galaxy. In some galaxies three supernovae have been detected in a space of twenty years, too high a rate to be a statistical fluctuation. It is suggested that there may be a class of galaxy in which the supernova rate is higher than average, i.e., $1/10-50$ years/galaxy. The rate of occurrence of supernovae is important both for radio astronomy and cosmic ray studies; a reliable figure will not be available until better statistics are obtained.

3.4 Classification of supernovae

Although all supernovae are superficially alike, there are sufficient differences to indicate that the two types are separate phenomena with apparently unrelated triggering mechanisms. The features that characterize the two types are summarized below:

(A) TYPE I

(i) *Spectra.* Despite intensive study no single feature of Type I supernovae spectra has been definitely established. The most notable features are broad emission bands in the region 3500 to 5000 Å. The same bands are seen in almost all supernovae of this type but the intensities and widths vary. The brightest band is centred at 4680 Å but its origin is not yet established. The spectra of supernovae are peculiar in the relative absence of hydrogen lines, indicating that the stars are mostly comprised of heavier elements and are near the end of their stellar evolution. The emission lines are seen against a continuum whose intensity is more variable than that of the lines. The Doppler width of the spectral lines suggests velocities of the order of 1–3000 km/sec.

(*ii*) *Light curve*. In contrast to their confused spectra, light curves of Type I supernovae are clear and reproducible. The general features are shown in Fig. 3.4. The sharp rise to maximum is followed by a slow decline; the duration of the period of maximum emission is typically 50 days. Subsequently the luminosity decays exponentially (i.e. a linear variation on a semi-log scale) with a half-life of about 50 to 70 days. This linear decline in m has led to the suggestion that the optical emission is associated with the radioactive decay of certain isotopes produced in the explosion which have half-lives of this order. The decay

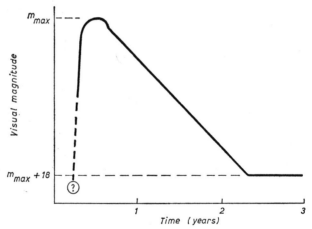

Fig. 3.4. Light curve of a Type I supernova

continues for 2–3 years at which point the luminosity reaches a steady value. There is no evidence of further disturbances. The overall change of magnitude during the outburst is greater than 15.

(*iii*) *Occurrence*. Type I supernovae occur in all types of galaxy. Within a particular galaxy they have a spherical distribution, characteristic of Population II stars, with which they are believed to be associated. Estimates of the rate of occurrence vary from 1/1000 years to 1/4000 years. Both Tycho's supernova and Kepler's supernova are believed to have been Type I but there is some doubt as to how the

Crab Nebula should be classified. All of these supernovae occurred within 1·5 kpc of the solar system. Since a sphere of radius 1·5 kpc is only 1% of the total volume of the Galaxy it is unlikely that all the supernovae in the Galaxy in the last 900 years occurred in this small volume. If the rate in this volume element was typical of the whole Galaxy, then the total rate is given by:

$$R(\text{Galaxy}) \sim R(\text{volume element}) \times \frac{\text{volume of Galaxy}}{(4/3)\pi(1\cdot5 \text{ kpc})^3}.$$

This gives a rate similar to that deduced for the galaxies with the highest supernovae rates.

(iv) *Energy.* Type I supernovae are old stars in the final stages of their evolution. Their masses are similar to that of the sun. The shell of gas ejected by the explosion contains about $10^{-1} M_\odot$ and the total energy liberated explosively is about 10^{48-49} ergs. At maximum the luminosity is typically two magnitudes brighter than Type II supernovae. Because of their brightness at maximum they are relatively easy to detect. Although they occur less frequently than Type II, most of the supernovae detected have been of this type.

(B) TYPE II

(i) *Spectra.* The spectral features are easily recognized and are quite similar to those found in novae. The Doppler broadening is considerable, suggesting velocities of about 7000 km/sec. Strong H_α and H_β lines are observed. From their spectra it appears that Type II supernovae are essentially the same phenomenon as novae, although on a larger scale.

(ii) *Light curve.* The variation of luminosity with time is irregular so that no characteristic decay time or pattern can be deduced.

(iii) *Occurrence.* Like Population I stars they occur in spiral galaxies preferably along the spiral arms. They do not occur in elliptical galaxies. Because of the small number observed, estimates of their rate vary from 1/30 years/galaxy to 1/400 years/galaxy.

(iv) *Energy.* The total energy radiation in the visible is 10^{50} ergs. When account is taken of radiation at other wavelengths, neutrino emission and kinetic energy imparted to the gas clouds, the total energy is about 10^{52} ergs. Type II supernovae are the result of the rapid evolution of massive young Population I stars. Although their optical luminosities are less than Type I, the mass of gas ejected (about $10 M_\odot$) is very much larger.

(C) OTHER TYPES

Zwicky (1965) has noted a small number of peculiar supernovae which do not appear to fit either of these categories. He proposes three other types, but since the numbers involved at this stage are small, it is difficult to isolate their characteristic features. It is not surprising that an event as energetic and explosive as a supernovae should show so many different features; from the theorist's point of view it is most important to isolate those features which regularly occur and are fundamental to the process.

TABLE 3.1

Comparison of novae and supernovae

FEATURE	N	SN I	SN II
Energy released in the explosion [ergs]	10^{44-45}	10^{47-48}	10^{51-52}
Absolute magnitude at maximum [m]	$-7 \rightarrow -8$	$-17 \rightarrow -18$	$-16 \rightarrow -17$
Mass ejected [$M_\odot$]	10^{-3}	10^{-1}	10
Velocity of gas shell [km/sec]	1000	3000	7000
Rate/year/galaxy	50	$2 \cdot 5 \times 10^{-3} \rightarrow 10^{-2}$	$2 \cdot 5 \times 10^{-3} \rightarrow 3 \times 10^{-1}$
Associated stellar types	Red dwarfs	Population II	Population I

3.5 Supernovae remnants

The identification of the strong radio source Taurus A with the Crab Nebula was the first optical identification of a radio source. Optical identification of radio sources now constitutes a major portion of optical and radio studies. The Crab Nebula was also the first x-ray source to be identified. It is thus one of the most prominent and important sources in the sky at all wavelengths.

Since the positions of SN 1572 and 1604 were known precisely from their discoverers' records, observations were made in the early 1950's to determine if their remnants were detectable radio emitters. The success of these observations has led to the supposition that all Galactic radio sources (with the obvious exception of H II gas clouds and flare stars) are supernovae remnants. As the remnants are not strong emitters at optical wavelengths, the technique has been to identify the remnant by its characteristic radio emission and subsequently to search for optical emission. Cassiopeia A is a typical case; although it is the brightest radio source in the sky, the supernovae outburst was not detected at its time of occurrence (believed to be less than 2000 years ago) and the only visible remnant is some very thin filaments. Another prominent source is the Veil Nebula in Cygnus, believed to be the remnant of an outburst 10,000 years ago. The characteristics of the principal remnants are summarized in Table 3.2 [Hayakawa (1963)].

TABLE 3.2

Principal supernovae remnants

SOURCE	AGE (years)	TYPE	DISTANCE (pc)	RADIO POWER (erg/sec)
Cassiopeia A	250	II	3400	3×10^{34}
Kepler SN 1604	360	I	1000	4×10^{31}
Tycho SN 1572	400	I	360	1×10^{31}
Crab Nebula	900	I?	1100	3×10^{33}
Veil Nebula	5×10^4	II	770	4×10^{32}

3.6 Crab Nebula

Although 900 years have elapsed since the explosion, the proximity of the remnant, plus the observations of the Chinese astronomers at the time of the outburst, make the Crab Nebula the prime object for investigation of the mechanism of the supernovae explosion. These studies are also relevant to the theory of cosmic ray origins, the latter stages of stellar evolution and thermonuclear processes in highly evolved objects. The high rate of energy release is, of course, the most intriguing aspect for high-energy astrophysics.

The Crab Nebula has been widely used as a basis for comparison of observation with supernovae models. This is somewhat unfortunate since the source is not really typical and hence may require a special sequence of events to explain it. This uniqueness is typified by the uncertainty as to whether the source is truly a Type I supernovae remnant.

Radio observations have been made over five octaves of frequency; the most important result has been that the radio spectrum is almost flat. The entire radio region can be represented by a power law of the form

$$I(\nu) = k\nu^{-\alpha}, \qquad (3.1)$$

where $\alpha = 0.27$ and k is a constant. Recently there has been some indication of a slight variation in this simple pattern at high frequencies. At all radio frequencies the angular size of the source is about 5 minutes of arc. The optical radiation from the filaments is in the form of spectral lines; the amorphous component radiates a continuous spectrum which can be represented by a power law

$$I(\nu) = k'\nu^{-1.1}.$$

The value of the exponent depends on the amount of interstellar absorption assumed. This radiation has been found to be 17% polarized. The radio spectrum power law, if extrapolated to higher frequencies, passes close to the optical continuum observations (Fig. 3.5),

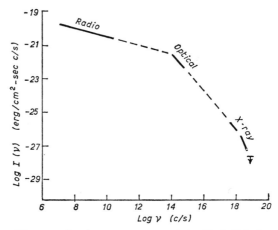

Fig. 3.5. Continuum spectrum of the Crab Nebula

suggesting that the entire continuum may have a common origin. This is confirmed by the observation of polarization at high radio frequencies.

That the Crab Nebula is not the passive remnant of a violent explosion is evidenced by the observation of continuing activity. Wisps of gas with velocities of the order of 10^{-1} c are observed in the central regions to undergo rapid changes; these point to the continuous injection of high-energy particles from some central source. The nature of the source is still a mystery; even a catastrophic explosion would be expected to leave some kind of core. Photographs show two small stars lying near the centre of the Nebula, but it is not clear whether they are actually in it or whether they merely lie along the line of sight (see Chapter 12).

If the core of the remnant is a neutron star, as has been suggested, then it may be invisible at optical wavelengths. The observation of x-rays was thought to be evidence of this, but later observations showed that the x-ray source had a finite size, roughly comparable to the amorphous optical source.

The flat radio spectrum is characteristic of thermal emission from an optically thin gas. Shklovsky (1960) has shown that if the radio spectrum from the Crab has this origin, then the filaments must have an electron density $N_e \sim 5 \times 10^4/cm^3$ and a temperature $T > 4 \times 10^6$ °K. From the optical observations, upper limits of $N_e < 2 \times 10^3/cm^3$ and $T < 1\cdot5 \times 10^5$ °K are deduced. In addition the spectrum should stay flat out to optical wavelengths, whereas a definite decrease is detected.

It was originally thought that the optical continuum was caused by free-free and free-bound transitions excited by the very hot central remnant of the explosion. This thermal origin proved to be unsatisfactory because of the excessive demands made on the central star, which, if this model were correct, would have several unusual features.

In the filaments the dominant lines are O II and N II. From the transition probabilities of these lines it is deduced that the temperature and density of the filaments is low. From the direction of the polarization vectors over the Nebula, the structure of the magnetic fields can be estimated. These fields show considerable variations over the Nebula and seem to be aligned along the filaments (Fig. 3.6). The magnetic fields probably define the outer boundary of the expanding gas shell which moves with velocities of 1000 km/sec. This velocity

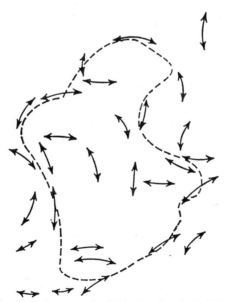

Fig. 3.6. Distribution of magnetic fields in the Crab Nebula relative to amorphous optical source

is such that if it has held since the outburst, the dimensions of the Nebula now would indicate that the explosion occurred 800, rather than 900, years ago. It is assumed therefore that some interaction between the gas shell and the magnetic fields has caused the former to accelerate above its original value. The mass of the filaments is estimated as $10^{-2} M_{\odot}$, considerably less than the mass normally assumed to be ejected from supernovae.

It is now accepted that the radio and optical continuum is due to magnetic bremsstrahlung radiation. This is confirmed by polarization measurements. Using values of $H \sim 10^{-4}$ gauss, the electron energies are $\sim 10^9$ eV for the radio emission and $\sim 10^{11-12}$ eV for the optical emission. The change in spectral index between radio and optical regions of the spectrum could be due to the lifetime of the optical-emitting electron being shorter than the lifetime of the Nebula. Although the mass in the form of relativistic electrons is negligible, the energy concentration is considerable. Ginzburg and Syrovatskii (1964), on the basis of equipartition arguments outlined in Chapter 2, derive the parameters listed in Table 3.3 for the Crab Nebula.

TABLE 3.3

Crab Nebula

Angular Size	5 minutes of arc
Field strength H	10^{-3} gauss
Total energy of relativistic particles E	$2 \cdot 7 \times 10^{48}$ ergs

3.7 Supernovae collapse theory

Many theories of supernovae collapse have been proposed: most of these have been qualitative and have described the basic mechanisms without reference to the details. The parameters are usually chosen to match the observational results so there is no possibility of rejection of a theory through confrontation with experiment. In these circumstances the choice of a supernovae theory is somewhat subjective, depending as much on the elegance of the proposals as on their physical consistency. The most detailed treatment of an explosion has been that by Hoyle and Fowler (1960, 1964), the results of which will be summarized below.

Since fusion processes are the dominant energy sources in stars, it is natural to consider first some fusion process as the source of the supernovae energy release. While fusion processes are essentially violent, requiring high temperatures and releasing more energy per nucleon than any other nuclear process, a star is normally so constituted that there is a high degree of mechanical stability. Only occasionally are instabilities seen and, even then, they are comparatively local and on a small scale compared with the total energy of the star. The ability of the layers of hot gas that constitute a star to rearrange, in the event of a local instability so as to preserve overall mechanical equilibrium, makes the occurrence of a catastrophic instability, such as occurs in a supernova, all the more unique. Only a process which releases very large energies in a time sufficiently short that a counteracting redistribution cannot take place is acceptable as a trigger for a supernovae outburst. It is estimated that for a catastrophic explosion about 10^{50} ergs must be released in less than 100 seconds.

The time scales associated with some of the most important stellar fusion reactions are shown in Table 3.4; also shown is the energy released per gram in the reaction. In no case can a model be devised which gives a chemical constitution such that the rate of energy release requirement can be satisfied. The fusion of light elements cannot therefore be considered as the basis of the supernovae energy release.

TABLE 3.4

Stellar fusion reactions

INGREDIENTS	ENERGY (per g)	TIME SCALE	NOTES
Protons	6×10^{18} ergs	Slow	High temperature required
Protons + light nuclei	10^{16} ergs	Slow	High density of light nuclei required
Carbon	5×10^{17} ergs	Fast	Unlikely composition

Hoyle and Fowler consider two cases: degenerate stars with $M \sim M_\odot$ and non-degenerate stars with $M \sim 30 M_\odot$. These two cases correspond to supernovae of Types I and II respectively. It is convenient to consider the latter case first.

The pre-supernovae star is taken to be near the end of its nuclear evolution; for a massive star this stage is quickly reached. If the central temperature is 5×10^9 °K the core will consist of iron and other heavier elements. This hot core will be surrounded by a shell of lighter nuclei at a lower temperature (Fig. 3.7). These lighter elements are the fuel for the explosion but the reaction rates of light elements at temperatures less than 10^9 °K are slow; only if a triggering process takes place in the core, which raises the temperature suddenly, will this energy be released explosively. Hoyle and Fowler propose that an implosion takes

D

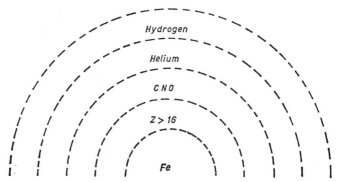

Fig. 3.7. Shells in star prior to supernova outburst

place in the core. The resulting kinetic energy of the collapsing gas shells will raise the temperature sufficiently for the lighter elements to fuse in a time of the order of one second.

The crucial feature in this process is the implosion which, Hoyle and Fowler show, will come about if heat can be absorbed in the core in a time short enough for it to be impossible for the star to rearrange itself without a complete breakdown of mechanical stability. The absorption process is the breakdown of the Fe^{56} nuclei in the core to helium nuclei and neutrons,

$$Fe^{56} \rightarrow 13He^4 + 4n.$$

This is an endoergic reaction which only takes place at the high temperatures and densities found in the stellar core. The reaction is the reverse of the normal fusion chains in which Fe^{56} nuclei are built up from protons with the release of energy. Since the fusion reactions can also take place an equilibrium state will exist in which half the total mass of the core will be iron, the other half helium and neutrons. The condition for this equilibrium is given by:

$$\log \rho = 11 \cdot 62 + 1 \cdot 5 \log T_9 - \frac{39 \cdot 17}{T_9}, \qquad (3.2)$$

where ρ = density in g/cm^2, and T_9 = temperature in units of 10^9 °K. This relation is plotted in Fig. 3.8. Since this is a plot of $\log \rho$ and T_9, a large change in density produces only a small change in temperature.

From equation (1.5), a star is stable if the gravitational energy is balanced by the thermal energy. When radiation pressure is significant the gravitational energy can exceed the thermal energy by a factor less than 2 and no implosion will take place. The star will evolve with gradual contraction and the nuclear fuel will burn to maintain the thermal energy. Consider a massive star which has evolved in this way to the point x in Fig. 3.8. Evolution will continue with the temperature

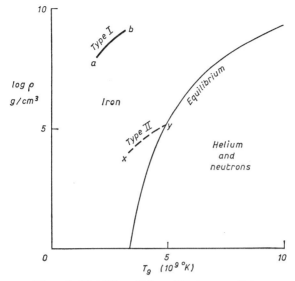

Fig. 3.8. The (He-n, Fe) equilibrium condition

increasing slowly with density until the point y is reached. The star will now tend to evolve along the equilibrium curve (eqn 3.2) since at this temperature an energy of about 2×10^{18} ergs/g is required to cross into the helium–neutron region. As the equilibrium curve is followed, the conditions for mechanical stability are no longer satisfied. The rapid increase in density causes an increase in gravitational energy. The corresponding increase in temperature is small, so that the increase in thermal energy is not sufficient to balance the forces tending toward collapse. Implosion takes place in the time of free fall (1 second) and the surrounding layers also collapse. The resulting high temperature causes the layer of light nuclei to burn explosively with the ejection of the outermost layers as an expanding gas envelope. The energy released is estimated as 2×10^{51} ergs.

Hoyle and Fowler estimate that this implosion will only take place if the stellar mass $M \geqslant 10\ M_\odot$. For stars with $M \sim M_\odot$ some other mechanism must be considered. Near the end of the evolution of these stars, most of the mass will be degenerate. When the central temperature has reached 2×10^9 °K, the star is essentially unstable (point a in Fig. 3.8), so that a small rise in temperature (point b) is sufficient to cause an explosion. This explosion is similar to the explosion that occurs in the more massive stars following implosion; in this case the inherent instability of degenerate matter is the triggering mechanism. 7×10^{50} ergs can be released explosively in this way.

In either of the cases considered above the star must undergo a non-explosive evolution until a temperature of 2×10^9 °K is reached. If $M < 1.16 \, M_\odot$, then the degeneracy pressure will be sufficient to prevent collapse until the star has cooled off to become a white dwarf. If $M > 1.5 \, M_\odot$, then mass will be ejected until a stable configuration is reached. Only where $M > 10 \, M_\odot$ will nuclear evolution proceed to the point of catastrophic implosion.

Several features of the Hoyle–Fowler theory are in excellent agreement with the observed data. (a) Type I supernovae do not exhibit the spectrum of hydrogen. If the star is small and at a late stage of nuclear evolution, not much hydrogen would be expected, even in the outer layers. On the other hand, although the centre of massive young stars may undergo a rapid nuclear evolution and use up its hydrogen supply, the outer layers will still consist of almost pure hydrogen. The emission spectrum of hydrogen is clearly visible in Type II supernovae. (b) In Type II supernovae a significant amount of the radioactive isotope Californium 254 may be produced. This has a half-life of 55 days, which is of the same order as the characteristic half-life of the light curve. It is not clear, however, just what the connection between the decay of this isotope and the light emission is, nor is the light curve half life always exactly 55 days.

These models have been extended by Hoyle and Fowler to take into account the energy loss by neutrino processes in the pre-supernova stage. They conclude that in massive stars the most important process is $e^- + e^+ \longrightarrow \nu + \bar{\nu}$. While this energy loss will tend to speed the evolution of the star, the energy loss will not compete with the reaction

$$Fe^{56} \longrightarrow 13He^4 + 4n.$$

Colgate and his collaborators (1965) have extended the Hoyle–Fowler treatment with particular emphasis on supernovae as sources of cosmic rays. Whereas Hoyle and Fowler have treated the explosion in terms of the nuclear processes, Colgate has computed the mechanics of the outburst in terms of the hydrodynamics of the star. Starting with a highly evolved star of mass $10 \, M_\odot$, the subsequent implosion and explosion are computed qualitatively. The equation of state for these calculations includes the effects of electron, nuclear and radiation pressure and the endoergic reactions $Fe^{56} \longrightarrow 13He^4 + 4n$ and $He^4 \longrightarrow 2p + 2n$. With only these conditions the implosion proceeds indefinitely with no force sufficient to balance the gravitational forces. The effects of angular momentum and the magnetic field were considered as possible forces for halting the collapse; in each case the resultant force was insufficient.

When the collapse has reached the stage that densities $\sim 10^{11}$ g/cm^3 exist, a neutron star may form. This stable configuration will halt the collapse and cause a shock wave to propagate outwards from the collapsed core, carrying with it a portion of the star's mass. At this stage the gravitational energy of the core (which constitutes 20% of the stellar mass) is about $2 - 4 \times 10^{20}$ erg/g; if only 10% of this is carried away by the shock wave, it would be sufficient to remove 80% of the stellar layers surrounding the core. Neutrino emission by inverse beta decay in the core will carry away a considerable amount of energy. The neutrinos will have energies of 12 MeV so that 10% of them will be absorbed by the outer layers of the star and will enhance the shock-wave effect.

For $M = 10\ M_\odot$ the energy emitted as ejected matter, that is, cosmic rays, is estimated as 10^{51} ergs (about 10^{-4} of the total stellar rest mass). This large energy may provide the solution to the cosmic-ray energy source enigma (see Chapter 4). In the shock-wave front electrons will receive additional acceleration from Compton scattering with photons from the radiation field. The resulting charge separation of the electrons and heavier positive ions can accelerate particles in the electrostatic field to energies of a few GeV per nucleon. Rapid nuclear synthesis behind the shock wave will cause a preponderance of heavy elements; this is in agreement with the observed chemical composition of the cosmic radiation. As the shock wave moves to regions of smaller densities, its velocity increases and, with it, the relativistic gas shell (cosmic rays). The resulting cosmic-ray energy spectrum will be a power law

$$N(>E) = kE^{-1.5},$$

which agrees with experiment. Rapid plasma oscillations may accelerate particles to 2×10^{19} eV; otherwise the upper limit to the cosmic ray energy is 3×10^{16} eV.

While the Hoyle–Fowler–Colgate treatments cannot at this stage be either verified or disproved, they do constitute a plausible mechanism which agrees with most of the observations. This does not mean that one of the many unorthodox theories that have been proposed will not ultimately prove to be more satisfactory. For the present the above treatment can be taken as a working hypothesis, particularly for branches of astrophysics which only incidentally involve supernovae.

4
Origin of the Cosmic Radiation

4.1 Introduction

The cosmic radiation is the most bewildering of the extra-terrestrial radiations, which provide the sole information channels between the surface of this planet and the rest of the universe. Its detailed composition, and particularly its high energies, have led to an intensive investigation over the last fifty years. This investigation has had a two-fold purpose: (1) to investigate the nature of the radiation and hence deduce its origin; (2) to use the radiation as a means of investigating high-energy nuclear interactions and cascade theory. The most spectacular results have come from (2) but, with the advent of particle accelerators, attention has shifted to the astrophysical aspects of the radiation.

Although the first effects of the cosmic radiation were noted at the turn of the century, it is only in the last twenty years that the various forms of the radiation have been identified. This has been largely because of the effect of the earth's atmosphere, which results in the interaction of the high-energy particles at high altitudes; the products of these interactions constitute the secondary radiation. The primary radiation can only be detected at balloon altitudes; recently satellites have permitted the first observations above the earth's atmosphere. The development of cosmic ray detection techniques has paralleled the development of nuclear particle detectors. In the thirties and forties the study of cosmic ray secondaries made important contributions to elementary particle physics. At low energies the instruments carried in balloons or satellites are similar to those used in particle accelerator experiments. At energies greater than 10^{13} eV the only feasible method of investigation (because of the low intensity) is to study the secondary particles at mountain altitudes or at sea level. Because of the low density of the atmosphere the secondary particles spread over a wide area; these extensive air showers can contain as many as 10^{10} particles spread over an area of the order of one square kilometre. The vertical thickness of the atmosphere is 27 radiation lengths (1030 g/cm^2); the particles at sea level are thus many generations removed from the primary (Fig. 4.1).

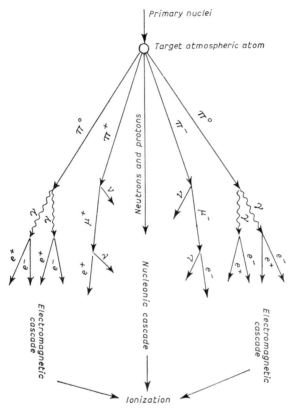

Fig. 4.1. Development of extensive air shower in the atmosphere

Extensive air shower arrays for the detection of showers with $E > 10^{16}$ eV are large and costly affairs involving hundreds of particle detectors. The particle detectors commonly used include Geiger counters, scintillation counters, Cherenkov detectors, cloud chambers, spark chambers, neon tube hodoscopes. These detectors are so arranged that they provide the following information about the shower: (*a*) spatial distribution of soft (electrons, photons) and hard (nuclear, mesonic) components, (*b*) direction of primary, (*c*) total number of particles and their energies and hence the energy of the primary.

Because of increasing interest in the cut-off of the primary particle spectrum (4.7) E.A.S. arrays have steadily increased in size. The first

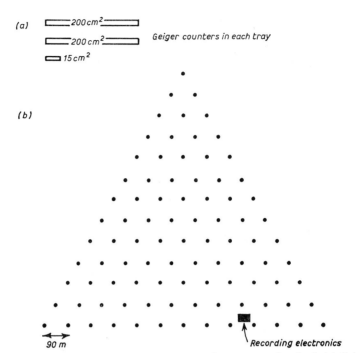

Fig. 4.2. Harwell E.A.S. array: 91 trays, covering an area of 0·5 km². (a) Geiger trays; (b) geometry of array

TABLE 4.1

Large air shower arrays

LOCATION	BASIC UNITS	OBJECTIVES AND SPECIAL FEATURES
Havarah Park, Great Britain. 230 m above sea level.	4 large detectors each consisting of 15 water Cherenkov counters. Area = 1 km². Energy > 10^{17} eV.	Directional anisotropies. Absorption of showers at large zenith angles.
Mount Chacaltaya, Bolivia. 5200 m above sea level.	17 large unshielded scintillators. Neon tube hodoscope. Cloud chamber. Energy > 10^{15} eV.	High altitude permits the detection of small showers at their maximum development. Primary gamma-ray directions. Heavy primaries. High-energy nuclear interactions.
Moscow University.	20 scintillation counters. Energy > 10^{16} eV.	Large underground μ meson detector. μ meson distribution.
Volcano Ranch, U.S.A. 2000 m.	20 scintillation counters. Energy > 10^{17} eV. Area = 8 km².	Energy spectrum. Directional anisotropies.
Sydney, Australia (proposed).	Units of two scintillation counters separated by 50 m, 1 km apart. Area = 250 km². Energy > 10^{18} eV.	Each unit autonomous. Data stored on tape. Time signal from central station. Energy spectrum. Directional anisotropies.

really big array was operated at Harwell between 1954 and 1958 (Fig. 4.2). The principal features of some of the largest existing arrays are summarized in Table 4.1. Even the giant Sydney array will have a counting rate for particles of 10^{21} eV (if they exist) of only one per three years. This experiment probably represents the largest feasible array using conventional techniques. A number of unconventional techniques have been pursued in the hope that the largest showers could be detected with modest equipment. These techniques include the use of radar, the detection of Cherenkov or fluorescent light in the earth's atmosphere generated by the passage of the secondary particles, or the detection of radio emission from the charged secondary particles. None of these can yet compete with the conventional arrays of particle detectors but they may be the best hope for the future, particularly if the primary spectrum gives no indication of a cut-off.

4.2 Experimental data

Before considering the conflicting theories of cosmic-ray origins, the principal experimental data on the nuclear component will be briefly reviewed under the following headings: chemical composition, energy spectrum, energy density, anisotropies.

(A) CHEMICAL COMPOSITION

The unusual chemical composition in the energy range 10^9 eV to 10^{12} eV has been investigated using emulsion stacks exposed near the top of the earth's atmosphere, usually in balloons. The unexpected result that the composition is radically different from the estimated universal distribution of elements is one of the most difficult features for any origin theory to explain. Since the data on individual elements are somewhat meagre, it is usual to classify the elements into groups of elements according to their atomic numbers. In Table 4.2 the cosmic

TABLE 4.2

Comparison of cosmic radiation and universal abundances

GROUP OF NUCLEI	z	COSMIC RADIATION (MEASURED)	UNIVERSE (ESTIMATED)
p	1	700	3000
α	2	50	300
L	3–5	1	10^{-5}
M	6–9	3	3
H	10–19	0·7	1
VH	20	0·3	0·06

Note: These figures have been normalized to (H + VH) = 1. This corresponds to an intensity of H + VH nuclei of $\sim$1·9 nucleons/m²/sterad/sec.

radiation and universal compositions of the various groups are compared; the H + VH groups are normalized to 1. These values are for
rigidity $R > 4.5$ GV, where the rigidity is defined as
$$R = p/Z,$$
where p is the momentum in GeV/c and Z is the atomic number.

The universal chemical abundance is the composite of data from
many disciplines. The abundances in the earth can be determined
from geological studies; for the heavier elements these abundances are
in agreement with stellar abundances. Because of the general agreement, the abundance of relatively rare elements on the earth can be
taken as representative of the universe. The earth is particularly poor
in hydrogen and helium; this absence is expected in view of the
volatile nature of these elements. From stellar studies these elements
are by far the most abundant; spectroscopic studies of the sun allow
qualitative estimates to be made. Of the other stars only the hot O, B
and A types are bright enough for an estimate to be made of their
compositions from the relative intensities of their spectral lines. The
general agreement in cosmic abundances from earth, meteor and
stellar data suggests that all the matter in the universe was well-mixed
at some stage.

The most significant feature in the comparison of the cosmic-ray and
universal distributions is the relatively large number of light nuclei
present in the former. This is particularly unexpected since the cross-
section for nuclear interactions for heavy nuclei is considerably greater
than for light nuclei. The presence of any heavy nuclei at all indicate
that the primaries have only traversed a density of matter of a few
grams per square centimetre.

The factor of 10^5 difference in the relative amounts of the L nuclei
indicates that their presence in the cosmic radiation is largely due to the
fragmentation of heavier nuclei. The detailed chemical composition
seems to confirm this hypothesis but there are some unusual features,
e.g. a high proportion of the He^3 nucleus is found but may be explain-
able by a higher universal concentration than was previously indicated.

A theory of fragmentation would indicate that most, if not all, of the
nuclei, when they leave the source are heavy with $Z \geqslant 20$. Recently a
surprisingly large number of very heavy nuclei, with $Z > 40$, have been
found; these observations have not yet received a satisfactory explanation.

Reliable data on fragmentation probabilities are not yet available
and it is difficult therefore to trace the primary nuclei at the earth back
to their source. Fortunately this is a field which can be studied in the
laboratory using particle accelerators or secondary cosmic radiation.

No reliable data are available on the chemical composition at higher energies. Up to 10^{14} eV, the ratio of p to α primaries indicates that the composition has not changed. Investigation of the composition at higher energies ($E \geqslant 10^{14}$ eV) is hampered by the sparsity of such particles. The only feasible method of investigation is to detect the resulting extensive air shower.

There is some evidence that in the energy region 10^{15} to 10^{16} eV, the proportion of heavy nuclei to protons increases. Above this energy this ratio quickly falls off and it is believed that above 10^{17} eV all the primaries are protons. These results are still tentative as the number of such high-energy showers detected is still comparatively small. This question is quite critical for cosmic-ray origin theories and will only be resolved by the extension of existing air shower techniques.

The presence of anti-particles in the primary radiation could be significant evidence of areas in the universe where large amounts of anti-matter are present; recent determinations show that the number of anti-protons present in the primary radiation is less than $0 \cdot 1 \%$ that of protons.

The term 'cosmic radiation' is usually taken to apply not only to relativistic nuclei but to all forms of high-energy radiation. Because of the short half-life of the neutron (12 minutes), neutrons can only be expected from comparatively local sources, e.g. the sun. Electrons were at one stage proposed as the principal component of the primary radiation. Recent measurements indicate that the electron component is only 10^{-2} that of the nuclear component of the same energy. The electron component is currently attracting much attention and is considered separately at the end of this chapter. The presence of gamma-rays and neutrinos in the cosmic radiation is expected, but measurements are difficult; these fields are considered in Chapters 9 and 10 respectively. Almost every strange particle proposed has, at some time, been proposed as a constituent of the cosmic radiation; these particles include the magnetic monopole and the quark. Thus far they have escaped detection here, as elsewhere.

(B) ENERGY SPECTRUM

Above 10^9 eV the energy spectrum of the primary radiation can be represented by a power law of the form $\mathcal{N}(>E) = kE^{-m}$, where $\mathcal{N}(>E)$ is the number of cosmic rays with energies greater than E, k is a constant and m is constant over limited regions of E. Between 10^9 eV and 10^{15} eV, $m = 1 \cdot 6$ but in the region 10^{15} to 10^{16} eV the value of m changes to $2 \cdot 2$. These steepened spectra are maintained till 10^{18} eV when m again becomes $1 \cdot 6$ (Fig. 4.3).

The change of exponent in the 10^{15} to 10^{16} eV region may arise from a number of causes. If the primary radiation comes from two different sources, each with a power law spectrum, then the resultant spectrum will not be a smooth power law, but will exhibit a change of exponent. The possibility that the change represents the transition from Galactic to extra-galactic sources has been proposed. Two sources or types of source within the Galaxy could also be responsible. Hoyle, in an effort

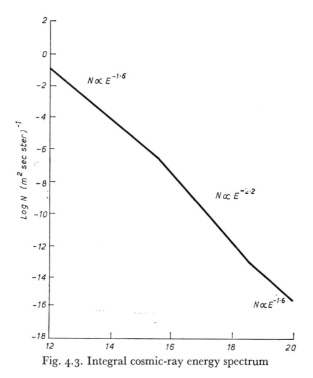

Fig. 4.3. Integral cosmic-ray energy spectrum

to resolve the presence of heavy nuclei with the passage through the dense interstellar space required by a confined Galactic theory, proposes that the primary radiation may be composed of two types, one of which has traversed the halo and the other the disc. The different histories of the nuclei could change both the chemical composition and the energy spectrum.

Certain features of the resulting air shower also change at $E = 10^{15}$ eV and it has been suggested that a change in the nature of the nuclear interactions might be responsible for the two effects. However, the persistence of the changed exponent over three decades of energy indicates something more fundamental.

The change at 10^{18} eV has been attributed to (i) a cosmological effect, (ii) a Galactic–extra-galactic contribution, (iii) a change in the composition of the primaries (see 4.7).

(c) ENERGY DENSITY

The total energy density of the cosmic radiation in the vicinity of the earth is about 10^{-12} ergs/cm³ which is of the same order as the Galactic starlight density. The recently discovered microwave flux, which is believed to be of extra-galactic origin, has a similar energy density. Due to the steepness of the frequency spectrum the majority of the energy comes from comparatively low-energy particles. The absence of a high-energy cut-off is hence of no consequence for the energy density.

Theories of cosmic-ray origins can be classified according to the region to which they assume the cosmic-ray density observed at the earth extends. Study of meteorites indicates that the radiation is not in any way local to the earth. Three regions are generally considered: these are classified in Table 4.3.

TABLE 4.3

Total cosmic-ray and rest mass energies

REGION	TYPICAL DIMENSION	VOLUME (cm³)	COSMIC-RAY ENERGY (ergs)	REST MASS (ergs)
Solar system	Radius of Pluto	10^{45}	10^{33}	3×10^{54}
Galaxy	Radius of Halo	5×10^{68}	5×10^{56}	3×10^{63}
Observable universe	Hubble distance	4×10^{84}	4×10^{72}	4×10^{76}

The absence of anisotropies, particularly at higher energies, eliminates the sun as the sole origin of the observed radiation. If the observed energy density is universal, then 10^{-4} of the total rest mass of the universe must be in the form of relativistic particles; such an energy distribution appears unlikely on a number of counts. If such a reservoir of energy is replenished over the time scale of the evolution of the Metagalaxy ($t \sim 3 \times 10^{17}$ sec) then sources with a power of 10^{55} ergs/sec are required. The age of the Galaxy is about 10^{10} years and there is evidence to suggest that the intensity of the cosmic radiation has not changed appreciably in that time. If this represents the lifetime of the particles in the Galaxy, then the rate of injection from the sources can be estimated. In fact the lifetime of particles is limited by nuclear absorption and diffusion out of the Galaxy; a reasonable estimate for the source power is therefore 3×10^{40} ergs/sec.

(D) SPATIAL ANISOTROPIES

Anisotropies are usually quoted in terms of the parameter

$$\delta = \frac{I_{max} - I_{min}}{I_{max} + I_{min}},$$

where I_{max} and I_{min} are respectively the maximum and minimum intensities observed in all directions. The experimental results obtained so far are only upper limits, and the results are quoted for a particular energy region. In Table 4.4 the upper limit for δ is expressed as a percentage.

TABLE 4.4

Spatial anisotropies

E (in eV)	δ
10^{14}	0·1
10^{16}	1
10^{17}	3
10^{18}	10
10^{19}	20

The lack of directional anisotropies is not so much evidence for an isotropic source distribution as for the existence of interstellar magnetic fields which successfully smear the original directions of the charged particles. Independent evidence for the existence of these fields comes from the Faraday rotation of linearly polarized radiation from extra-galactic radio sources and the magnetic bremsstrahlung radiation from relativistic electrons in the Galaxy. The value of the interstellar field is still in dispute and values from 10^{-5} to 10^{-7} gauss have been proposed. Even the lower limit of this field is sufficient to destroy the anisotropies expected in the high-energy flux from comparatively close sources. Consider the passage of 10^{14} eV protons from the Crab Nebula to the earth, a distance of 3×10^{21} cm. Taking $H = 10^{-7}$ gauss, the radius of curvature of the particle r, is given by

$$r = \frac{E}{300H} = 5 \times 10^{18} \text{ cm.}$$

$\ll D$, the distance to the source.

Thus the original direction would be completely obscured.

At sufficiently high energies, particles should retain enough of their original direction to provide a detectable anisotropy. It is unfortunate that it is in this high-energy region that statistical information is poorest. A number of point sources have been proposed as cosmic-ray sources. In addition there are a number of general directions in which a higher flux might be expected; these include the Galactic plane and the Galactic centre. The interstellar magnetic field appears to be ordered along the spiral arms; hence the spiral arm on which the solar system lies is another likely direction.

The absence of any such anisotropies in surveys using conventional extensive air shower arrays is evidence that the particles travel through the interstellar gas primarily by diffusion with magnetized interstellar gas clouds as scattering centres. Hence more precise values of δ might be expected to provide information on the distribution and order of magnetic fields in interstellar space.

4.3 Acceleration mechanisms

The most cunningly devised particle accelerator yet built by man can accelerate protons to energies of $\sim 10^{11}$ eV: the highest energy present in the cosmic radiation is as yet undetermined but is certainly at least 10^{20} eV. How such high energies are achieved in an apparently chaotic universe has been one of the most puzzling features of the radiation since its discovery. While the vast scale of the universe might be expected to be some compensation for the lack of organization, some efficient mechanism must be at work.

Laboratory accelerators invariably depend on electromagnetic acceleration either by the passage of charged particles through electrostatic fields or through non-static magnetic fields. It can be easily shown that the conductivity of inter-galactic or Galactic space is too high to permit electrostatic fields to endure long enough to produce effective acceleration. However, magnetic fields are closely linked to the particles; they thus have permanency in time but undergo spatial variations. A number of mechanisms have been proposed based on these non-static magnetic fields but the most generally accepted is that of Fermi or some variant of it. Fermi regards the acceleration as a statistical process; each region of magnetic field can be regarded as analogous to a large mass with which the charged particles are constantly in collision. This mechanism can be considered to operate in various regions, e.g., supernovae shells, interstellar space, inter-galactic space. The original Fermi theory was concerned with acceleration in interstellar space where the turbulent motions of the gas clouds (and hence of the magnetic fields) was estimated as $\sim$30 km/sec. In a collision a particle may gain or lose energy; on the average it will gain since a head-on collision is more probable than an overtaking collision. The impressive feature of the Fermi theory is that it leads naturally to a power-law spectrum. For a relativistic particle the gain in energy is proportional to the original energy:

$$dE/dt = aE, \quad \text{where } a \text{ is a constant}$$

therefore

$$E = E_0\, e^{at}.$$

E_0 is the value of E at $t = 0$. To find the observed energy spectrum the loss mechanisms must also be taken into account. The most important of these is ionization loss, which is given by

$$-(dE/dt) \propto 2 \text{ MeV g}^{-1} \text{ cm}^{-2}.$$

Once the particle energy is such that the exponential energy gain exceeds the ionization loss, the particle will be accelerated until it either leaves the accelerating region or suffers a nuclear collision. The lifetime T of the particle in the accelerating region is thus finite. The probability of finding a particle of age between t and $t + dt$ is given by

$$p(t)\, dt = \frac{1}{T} e^{-t/T}\, dt.$$

But

$$dt = dE/aE$$

therefore

$$t = \frac{1}{a} \ln (E/E_0).$$

The number of particles $N(E)\, dE$ with energy between E and $E + dE$ is proportional to the number of particles with age between t and $t + d$

therefore

$$N(E)\, dE \propto \frac{1}{T} e^{-(1/aT)\ln (E/E_0)} \frac{dE}{aE}$$

$$\propto E^{-(1+1/aT)}\, dE$$

$$\propto E^{-m}\, dE,$$

where

$$m = 1 + 1/aT$$

is a constant which depends only on the lifetime of the particles and the efficiency of the Fermi collision energy transfer.

Several variations of the Fermi theory have been proposed in which the exponent m has the value observed experimentally at the earth. The original suggestion that the particles are accelerated in interstellar space has now been rejected. A more detailed treatment shows that the original estimate of a was too high and that interstellar space is only capable of acceleration by a factor of ten at most.

Most theories now consider acceleration as taking place in the more confined region of a 'source'. Here an unusual chemical composition is easier to explain. The role of the interstellar space then is that of a diffusive medium; some acceleration may take place but the most important effect is that of smearing the original directions to give an isotropic flux. In addition the energy spectrum of the particles from different sources will also be smeared to give a power law with constant exponent. The net result has been to transfer the problem of acceleration from interstellar space to that of individual sources. Several mechanisms have been suggested to act in sources. The Extra-galactic theory is essentially a revision of the Fermi theory but on a Metagalactic, rather than galactic, scale.

4.4 The Galactic Origin Theory

The small amount of experimental data and the wealth of phenomena observed have led to a bewildering assortment of theories of cosmic-ray origins. Amid this complex of qualitative, and often contradictory theories, two broad schools of thought may be seen to emerge. The first, and probably the currently orthodox, theory is that of Galactic origin: the principal proponents of this theory are the Russian astrophysicists, Ginzburg and Syrovatskii (1964) and the radio-astronomer Shklovsky (1960). The Galactic theory proposes that all the cosmic radiation observed on the earth is of Galactic origin (some contribution from the sun is accepted for the lower-energy region). Cosmic rays may diffuse out of the Galaxy, but extra-galactic space is essentially devoid of cosmic radiation, so that this diffusion is a loss process.

The opposing view is less well formulated, but more generally accepted, in the Western world. The Extra-galactic theory considers that, while the low-energy region is of solar origin and the medium-energy region of Galactic origin, particles with $E \geqslant 10^{15}$–10^{16} eV must be considered to be of extra-galactic origin. The inter-galactic flux is hence considerable and contains a significant portion of the energy in the universe. Several versions of this hierarchical theory have been proposed, varying in the region in which it is considered that the extra-galactic radiation extends. The extreme view is that the observed energy density is universal and even the medium-energy cosmic rays are of extra-galactic origin. The recent discovery of quasars and quasi-stellar galaxies seems to favour the role of extra-galactic sources, in that the number of objects displaying violent phenomena on a galactic scale are increased.

(A) ENERGY SOURCES

The first requirement of a Galactic theory of origin is to show that there

E

are sources available within the Galaxy capable of providing the required power, i.e. 3×10^{40} ergs/sec. It is then necessary to account for the conversion of this energy to high-energy particles and to explain their unusual chemical composition.

A number of suggested sources have been reviewed by Ginzburg and Syrovatskii in terms of their available power.

(i) *Stable stars*. Low-energy cosmic rays have been observed from the sun in solar flares. However, optimistic assumptions about the energy output of stars such as the sun give only 10^{24} ergs/sec. There are an estimated 10^{11} stars in the Galaxy, corresponding to a total power of 10^{35} ergs/sec, well below the required figure. In addition the energy spectrum of the solar cosmic rays has an exponent of 5 and is composed predominantly of protons.

Giant stars have a luminous output of 3×10^{43} ergs/sec. They also possess highly active sub-photospheric convection zones and hence considerable free energy at their surfaces. However, while such stars could provide the energy density, no mechanism for the acceleration of particles in them has been proposed. They may be important contributors to the low-energy cosmic-ray flux.

It has also been proposed that the class of stars known as magnetic variables should be considered as cosmic-ray injectors. Since the sun has a maximum magnetic field of a few gauss in the areas of the photosphere, where solar flares occur, the magnetic stars have a magnetic energy at least 10^6 times that of the sun. If this factor corresponded to their improved rate of cosmic-ray production then their power would be $\sim 10^{30}$ ergs/sec.

The number of these stars is still undetermined; other spectral types also exhibit magnetic fields but generally of a smaller order. An upper limit would be that they comprise 1% of the Galactic stellar population, giving a total power of 10^{39} ergs/sec. This figure must be considered optimistic.

(ii) *Supernovae and novae*. One of the first suggestions of a cosmic-ray source was that of supernovaé, known to be amongst the most violently explosive and energetic objects in the universe. The subsequent discovery that the strong non-thermal radio emission could be explained in terms of magnetic bremsstrahlung of relativistic electrons in weak magnetic fields was confirmation of the association of these objects with high-energy particles. In the previous chapter the energy in the relativistic particles in the Crab Nebula was estimated as $2 \cdot 7 \times 10^{48}$ ergs. Using similar methods the total energy in other supernovae can be estimated; some of these are listed in Table 4.5.

A total energy of 10^{49} ergs per explosion is probably typical. To

TABLE 4.5

Energy in supernovae remnants

SOURCE	MAGNETIC FIELD (gauss)	U (ergs)
Cassiopeia A	10^{-3}	7×10^{49}
Cygnus filament	3×10^{-5}	2×10^{49}
SN 1604 (Kepler)	7×10^{-4}	6×10^{46}

estimate the total power it is necessary to know the frequency of occurrence. Taking a rate of 1 per 50 years gives a power of 10^{40} ergs/sec. This satisfies the injection power requirement to the order of the approximations used in these estimates.

In Colgate's model of supernovae explosions a total energy of 10^{51} ergs is predicted for a single Type II supernova. These large Type II supernovae are required every 1000 years to explain the observed energy density. Using such a model, protons could be accelerated to 10^{19} eV; extra-large supernovae, which are most likely in the Galactic core, could supply protons with $E = 10^{20}$ eV.

Hoyle has proposed a model somewhat similar to this, but the supernovae only inject particles into a local high concentration of supernovae where acceleration takes place. The peculiar feature of this theory is that it envisages a second shock wave which occurs in the remnant of the star about a year after the initial explosion. This time interval is necessary since the density of matter at the time of an explosion would be too great to permit the acceleration of heavy nuclei to relativistic energies. Rapid nuclear cooling by neutrino emission causes the second shock wave to propagate at a greater speed than the first and in a short enough period to cause it to catch up with the matter being carried along by the first.

Although there is not a continuous range of explosive objects stretching from supernovae to novae, it appears that similar phenomena are exhibited in each, although on a vastly different scale. In a novae explosion 10^{45} ergs are emitted in the visible part of the spectrum; it is reasonable therefore to take the cosmic-ray emission as 10^{45} to 10^{46} ergs. Taking the frequency of novae explosions as 100/year, the injected cosmic-ray power is 3×10^{39} to 3×10^{40} ergs/sec. This also satisfies the power requirement and it appears that if supernovae are responsible for a considerable proportion of cosmic rays, then novae must also make a significant contribution, possibly at lower energies.

(B) INTERSTELLAR SPACE AND THE CHEMICAL COMPOSITION

An important feature of the Galactic theory of Ginzburg and Syrovatskii is its ability to account for the observed chemical composition. Since the exact nature of the sources and their spatial distribution are unknown, it is difficult to estimate the amount of matter traversed by the cosmic rays; although it is assumed that the particles are localized within the Galactic halo, the nature of the boundary is unknown. Ginzburg and Syrovatskii have taken eight models in which various possibilities are considered. Among the sources considered are a point source at the Galactic centre, a homogeneous distribution in the Galaxy, extra-galactic sources. Boundary conditions vary from total reflection to free departure. In most cases it is assumed that motion through interstellar space is by diffusion.

Fragmentation probabilities are not sufficiently well-known to enable one set of parameters to be chosen; therefore in each of the eight models, four sets of fragmentation probabilities were used. Models were judged on their ability to give the observed chemical composition where the initial flux included no L group elements.

The best agreement was found for a model which assumed that all the sources were near the Galactic centre and that cosmic rays could diffuse freely out of the Galaxy. The total amount of matter traversed is 10 g/cm² ; from the relative concentration of matter in stars, dust and gas, it appears that most of this will be interstellar gas. The diffusion coefficient comes out at 10^{29} cm²/sec in agreement with other estimates.

In an extension of this model Ginzburg and Syrovatskii consider that the original cosmic rays consist of only heavy nuclei. Then, with the above model, they calculate the chemical composition expected at the earth; their values are in remarkable agreement with the observed values.

Recent observations have cast doubt on the existence of the Galactic halo. If the concept of the halo has to be disregarded, then the Galactic theory of cosmic-ray origins, which at the moment is well supported, will require revision.

4.5 The Extra-galactic theory

(A) THE LOCAL GROUP THEORY

As indicated previously the assumption that the observed cosmic-ray density pervades all space demands a disproportionate amount of the total energy of the universe to be in the form of relativistic particles. The Galactic theory avoids this difficulty by assuming the density is local; the inter-galactic density of cosmic rays is negligible by comparison and the total energy content therefore small.

Various theories have been proposed which extend the region of high cosmic-ray concentration beyond the Galactic boundaries; in general these are scaled-up versions of the Galactic theory. The first of these is the Local Group theory which has been developed by Sciama: the Local Group is the name given to the small cluster of galaxies to which the Galaxy belongs. This group comprises between 15 and 20 galaxies but the total galactic mass is only seven times that of the Galaxy.

The expanded galactic model proposed by Sciama regards the volume of the Local Group (4×10^{72} cm^3) as filled with cosmic rays which are generated in the galaxies. The density of inter-galactic gas is required to be 5×10^{-28} g/cm^3, which is 50 times the estimated inter-galactic gas density. It is necessary to assume that $U_c \gg U_H$ and that the magnetic field is strongly tied to the galaxies and effectively confines the motion of the cosmic rays. The accelerating sources are again taken to be supernovae and the number of supernovae is assumed to be proportional to the galactic mass.

Even making a number of assumptions for which there is no experimental evidence, this model fails to give the observed energy density by a factor of 10^2; however, the absence of a Galactic halo might require such a model which successfully predicts a path length of 5 g/cm^2 and explains the chemical composition.

(B) THE BURBIDGE–HOYLE MODEL

Burbidge first considered a limited extra-galactic region, that of the Virgo super-cluster, whose dimensions are ~ 30 Mpc and which is estimated to contain 1000 galaxies. The predicted energy density was 4×10^{-13} ergs/cm^3.

However, the discovery of radio galaxies, whose energy content is considerably greater than the ordinary galaxy, had the same effect on extra-galactic theories as that of supernovae had on the Galactic theory. In an object such as Cygnus A the total energy in the form of relativistic electrons is $\sim 10^{59}$ ergs. If it is assumed that $U_c = kU_e$ where $k = 10^2$, then $U_c = 10^{61}$ ergs. Such large energies are extremely difficult to explain; a survey of possible mechanisms by Burbidge indicates that the gravitational energy associated with a large system is the most likely source. The release of energy by the gravitational collapse of such an object exceeds that from thermonuclear reactions, previously considered the most efficient mechanism, by a factor of 10^2. The detailed mechanism of collapse has not been worked out and seems to meet with some difficulties. Whatever the mechanism, if such large energies are involved, the injection into inter-galactic space of particles with $E \sim 10^9$ eV is not unlikely.

Burbidge and Hoyle consider a region with radius 300 Mpc which is estimated to contain 10^3 radio galaxies. The lifetime of a radio galaxy is probably only 10^6 years, so that in 10^{10} years there may have been 10^7 such sources. A cosmic-ray energy of 10^{60} to 10^{62} ergs/source gives a density of 10^{-14} to 10^{-12} ergs/cm^3. The upper limit of source power would thus provide the energy density required; if the lower part of the energy spectrum came only from this limited region and the higher energies pervaded all space even the lower limit would be satisfactory.

This theory regards the radio galaxies purely as injectors with acceleration taking place in inter-galactic space by a Fermi-type mechanism. An explosive event within the source is rejected on the grounds that it fails to explain the presence of heavy nuclei. The explosion does, however, cause the injection into inter-galactic space of a large energetic mass, which acts like a piston on the cosmic rays already present and produces acceleration. Provided the expansion is rapid compared with the expansion of the universe, it is not necessary to consider the acceleration confined to a particular region.

As usual, the energy spectrum of the resulting radiation depends on the rate of acceleration and the time spent in the accelerating region. The value of the exponent comes out in good agreement with experimental values. In addition there is no difficulty in accounting for acceleration of particles to $E \sim 10^{20}$ eV; a marked change in spectrum is predicted above this value.

(c) OBJECTIONS

Present knowledge of extra-galactic conditions is extremely scanty. Any theory proposing the regions outside the boundaries of the Galaxy as the source of any phenomenon must be largely speculative, even by astrophysical standards. Such theories are closely tied to cosmology and the proponents have usually had to take a definite stance between the evolutionary, oscillating or steady-state theories. In general those who have put forward the extra-galactic theory have been members of the California–Cambridge steady-state school. Recent evidence seems to be against the steady-state model; the radio source counts point to an evolving universe and the failure of gamma-ray astronomers to observe annihilation radiation means that the continuous creation must take place in some unusual form. The only evidence for the steady-state theory seems to be the continuing evolution of galaxies. Current astrophysical opinion is in favour of the oscillating universe model.

Ginzburg and Syrovatskii (1964) have noted a number of arguments which point to the rejection of an Extra-galactic theory. Since radio galaxies are as yet not understood, the hypothesis that they are the actual sources of cosmic rays must be more speculative than that of the supernovae hypothesis; the latter have been studied in detail and the theory of implosion, following nuclear fuel exhaustion, seems generally accepted.

If the motion of the cosmic rays in the inter-galactic medium is by diffusion, then Ginzburg and Syrovatskii calculate that the only cosmic rays reaching the Galaxy since its formation are those generated within a radius of 50 Mpc. On the basis of equipartition arguments and the estimated upper limit of the inter-galactic magnetic field (from radio astronomy observations) they conclude that a cosmic-ray density of 10^{-12} ergs/cm^2 would be two orders of magnitude greater than the magnetic field density. In addition the density of the gas is estimated at 10^{-29} g/cm^3, giving a kinetic energy density four orders of magnitude less than the required cosmic-ray density. They conclude that the best estimate of the inter-galactic cosmic-ray density is 10^{-15} to 10^{-16} ergs/cm^3. Their estimated upper limit to the density expected, if the cosmic radiation was the remnant of an initial explosion, is 10^{-14} ergs/cm^3.

Since the inter-galactic gas density is much less than the Galactic gas density, losses by ionization are less serious. This not only means that particles can be injected at lower energies, but also leads to a large amount of energy being concentrated in low-energy particles; thus even if the universal cosmic-ray density is 10^{-14} ergs/cm^3 some of this energy may be in the form of unobservable sub-relativistic cosmic particles.

(D) NEW EVIDENCE FOR EXTRA-GALACTIC COSMIC-RAY SOURCES

The discovery of quasi-stellar radio sources (Chapter 6) must be taken as fresh evidence for an extra-galactic cosmic-ray origin. In a previous section the total energy of some radio galaxies was given. The radio luminosities of these objects range from 10^{38} to 10^{45} ergs/sec. The quasars, of which 200 are now identified, have radio luminosities of 10^{44} ergs/sec and optical luminosities of 10^{46} ergs/sec. It appears therefore that radio galaxies and quasars are in the same energy class. The basic difference is that whereas radio galaxies are extended objects, quasars appear to have sizes not far in excess of stars. If the time scale associated with them is the same as that estimated for radio galaxies, $\sim 10^6$ years, then the total energy emitted is in the region 10^{60} to 10^{62} ergs. Even on an idealized nuclear model, where the efficiency of mass-energy conversion is 0·01, this would require a total rest mass of $10^{10} M_{\odot}$.

It is obvious that the mechanism proposed for quasars must operate on a scale never before considered. In the last few years numerous theories have been proposed, all of which are satisfactory in some respect but fail to give a complete picture. In general the implications of these theories to cosmic-ray origins have not been fully worked out. However, the existence of extra-galactic objects, in which processes are occurring on a scale far beyond anything found in the Galaxy, must lead naturally to a revival of the Extra-galactic theory, particularly for the highest energy particles.

4.6 Cosmic electrons

(A) SPECIAL FEATURES

The electronic component of the cosmic radiation is one of its most interesting features. Although the ratio of electrons to protons of the same energy is about 10^{-2}, direct measurements were not possible until this decade when high-altitude balloons allowed instruments to be carried to within a few grams of the top of the atmosphere. The existence of these electrons had previously been inferred from radio observations. Primary cosmic electrons have several features which distinguish them from primary cosmic nucleons:

(i) Because of their smaller mass, they lose their energy more rapidly. Their lifetimes are short and they must be of Galactic origin.

(ii) Taken with the radio observations, the electron spectrum permits the strength and distribution of the interstellar magnetic fields to be estimated.

(iii) The electron component is more liable to solar effects; the variation of the electron flux with time can be directly related to solar modulation.

(iv) Because the electron component contains both positive and negative electrons, the origin of the radiation can be inferred. In particular, whether the electrons are directly accelerated or whether they are the by-product of the nuclear component can be determined.

(B) DETECTION TECHNIQUES

The fundamental problem in balloon-borne experiments is to distinguish the electrons from the more numerous low-energy nucleonic cosmic rays. The earliest experiments used cloud chambers or complicated combinations of particle detectors which allowed the unique identification of the electron. In general the range and energy loss of the particle was measured. The large flux of secondary electrons produced in the atmosphere or in the instrument supports complicates all these measurements. It is usually necessary to make measurements at a few altitudes to allow extrapolation to zero atmosphere. Fortunately most of the range of energies under investigation is within the range of particle accelerators so that accurate pre-flight calibrations can be made. At high energies ($>$10 GeV) nuclear emulsions are the most feasible detection technique; since the areas available are low, the number of events detected is of necessity small. At even higher energies electromagnetic cascades will be produced in the atmosphere, but these are indistinguishable from those caused by primary gamma-rays. Apart from one Russian satellite experiment, all the experiments on electrons, with $E > 200$ MeV, have been balloon-borne.

(c) EXPERIMENTAL RESULTS

Primary electrons were first detected by Earl (1961) and Meyer and Vogt (1961). Since then groups in the United States, Russia, India and Europe have made a number of measurements at energies ranging from 200 MeV to 200 GeV. Results fall under two headings: (i) energy spectrum (either differential or integral), (ii) charge ratio defined as

$$R = \frac{N(e^+)}{N(e^+) + N(e^-)}.$$

The differential electron spectrum is now well-known and can be represented by a power law with an exponent that increases with energy. Some typical values of exponent are given in Table 4.6 for various overlapping energy ranges.

TABLE 4.6

Differential flux measurements

ENERGY RANGE (GeV)	EXPONENT OF DIFFERENTIAL SPECTRUM
0·77 to 3·6	2·0 ± 0·5
1·85 to 15	2·2 ± 0·3
3 to 300	2·62 ± 0·05

Measurements of R are difficult, involving either artificially created magnetic fields or using the East–West effect of the earth's field. The experimental values for R are shown in Table 4.7. The last value listed was obtained in an emulsion experimen t [Daniel (1967)]; the lower

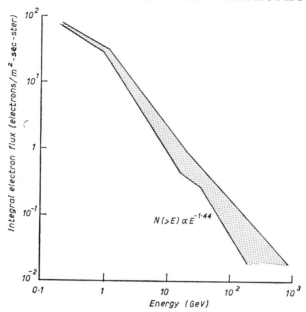

Fig. 4.4. Integral spectrum of primary electrons

energy values were obtained using an arrangement of Cherenkov and scintillation detectors and spark chambers with a permanent magnet [Hartman (1967)].

TABLE 4.7

Electron–positron ratio

ENERGY RANGE (GeV)	R
0·2 to 0·5	0·35 ± 0·10
1 to 2	0·11 ± 0·02
2 to 5	0·086 ± 0·02
5 to 10	0·098 ± 0·04
15 to 50	0·70 ± 0·20

(D) INTERPRETATION

Even with the data above, it is still not possible to deduce a definite theory of the origin of the electron component. Felten and Morrison (1966) have estimated the electron spectrum in the Galactic halo on the basis of known physical processes. This treatment will be briefly outlined. The chaotic field in the halo traps the electrons. Although the processes by which electrons can lose energy are relatively well-known, the production processes are more uncertain. Two possibilities are considered: (a) the electrons are produced as the result of proton–proton collisions in the Galaxy; (b) the electrons are accelerated directly. In either case it is assumed that the electrons are produced with a power-law spectrum of the form

$$\mathcal{N}(E).dE = \mathcal{N}_0 E^{-m}.dE.$$

Since it is unlikely that the electrons have a common origin, this assumption is a serious limitation on the treatment.

The mean energy loss per electron $\overline{dE/dt}$ for electrons of energy E is given by

$$-\frac{\overline{dE}}{dt} \sim a + bE + cE^2,$$

where a, b, c are functions of astrophysical and nuclear parameters. The processes which contribute to the various terms are shown in Table 4.8. The resulting spectrum will be a power law whose exponent depends on which of the above processes is dominant. The approximate value of the exponent is given in the third column of Table 4.8. The

TABLE 4.8

Electron energy loss processes

TERM	PROCESS	EXPONENT	ENERGY RANGE
a	Ionization	$m+1$	<10 MeV
bE	Diffusion	m	10 MeV–10 GeV
cE^2	Compton scattering Magnetic bremsstrahlung	$m-1$	>10 GeV

fourth column gives the energy range over which Felten and Morrison estimate the process indicated dominates; these values are for the Galactic halo.

The central region (10 Mev to 10 GeV) corresponds to those electrons which radiate the bulk of the non-thermal radio emission of the halo in the range 10 to 400 Mc/s where the magnetic field is taken as 2×10^{-6} gauss. The electron energy flux required to give the observed radio intensity has the differential spectrum

$$\mathcal{N}(E).dE = 80E^{-2\cdot4}/\text{m}^2 \text{ sec ster GeV}.$$

The measured differential spectrum between 2 and 15 GeV is in fair agreement, having the form

$$\mathcal{N}(E).dE = (46 \pm 18)E^{-(2\cdot2\pm0\cdot3)}/\text{m}^2 \text{ sec ster GeV}.$$

At lower energies the measured spectrum is considerably lower than that inferred from the radio background. This implies that because of a solar modulation effect the electron density at the earth is less than that outside the solar system.

Some indication of the electron sources can be deduced from the measured values of R. Electrons arising from the decay of mesons produced in proton–proton collisions should give a significant excess of positrons at low energies where the meson multiplicities are low. Hence at low energies R should be greater than 0·5. The observed values disagree with this prediction and point to direct acceleration for most of the observed electrons whose origin is unknown. The intensity of the positron flux is in agreement with that expected from proton–proton collisions in the Galaxy.

4.7 Upper limit to the cosmic-ray energy spectrum

With the experimental techniques in existence there is no evidence yet for a cut-off in the cosmic-ray energy spectrum. The highest energies are of particular interest, not only because of the tremendous concentration of energy in a single particle, but because several effects may become obvious which are peculiar to these high energies and because the energies are large enough to overcome the dispersive effect of inter-galactic magnetic fields.

Greisen (1965) has discussed some of these effects, with reference to the extension of air shower techniques to the 10^{21} eV energy region over the next decade. At 10^{20} eV the lifetime of a neutron is extended sufficiently for a neutron to reach the solar system from a distance of 1 Mpc. These neutrons might originate in the disintegration of heavy nuclei by inter-galactic radio photons. A neutron-induced shower would have many features similar to a proton-induced shower.

Both types of shower would be somewhat unique in their development at these high energies; time dilation would extend π^0 lifetimes so that γ emission would take place a couple of interaction lengths later than usual. Bremsstrahlung and pair production would also be inhibited so that showers would develop rather lower in the atmosphere than their lower-energy counterpart.

A fundamental limitation to the energies of cosmic rays is the size and energies of the sources. In a cyclotron or similar well-organized accelerator, the radius of the machine, R is of the order of the radius of curvature, ρ of the particle of energy, E in the magnetic field, H. In a cosmic source

$$R > \rho \quad \text{where} \quad \rho = E\beta/He$$
$$= 30\rho, \text{ say.}$$

If the magnetic field is too large, then energy loss by proton magnetic bremsstrahlung radiation will be excessive. Greisen deduces that the limiting condition for acceleration of 10^{21} eV protons is $H < 10^{21}\gamma^{-2}$ gauss where $\gamma =$ Lorentz factor of the proton or $R > 10^{-13}\gamma^3$ cm. The total magnetic energy $U_H = (H^2/4\pi)V_0$ where $V_0 =$ volume of the source. Assuming equipartition, then the total energy of the source $U \sim U_H$. Therefore from the above inequalities

$$U > 300\gamma^5 \text{ ergs.}$$

This sharp dependence of the total energy on γ shows that even the largest radio galaxies ($U \sim 10^{62}$ ergs) could only accelerate particles to 10^{21} eV. Greisen points out that should higher-energy particles be detected, then a degree of organization is indicated on a scale hitherto unknown in an astrophysical source. This upper limit to the spectrum is thus of considerable cosmological importance.

The recently discovered microwave background which seems to correspond to the universal 3 °K black-body field predicted as the remnant of a primeval explosion, imposes more severe limits on the high-energy limit of cosmic rays. The effect of this microwave flux on the cosmic-ray proton flux was first considered by Hoyle. Presuming that the upper end of the cosmic-ray spectrum is extra-galactic in origin he showed that the rate of energy loss by protons by the Compton effect is less than the loss rate due to the expansion of the universe. Greisen (1966) and Zatsepin and Kuzmin (1966) considered two other effects: pion production and pair production.

Greisen considers the microwave flux to have a photon density oı 550 photons/cm³ with a mean photon energy of 7.0×10^{-4} eV. The proton threshold for pion production by photons of this energy is 10^{20} eV. The cross-section is 400 μ barns at the $3/2$, $3/2$ resonance (2.3×10^{20} eV) with an average value of 200 μ barns. The mean path for interaction is 9×10^{24} cm; the energy loss per interaction is 10 to 20%, so for particles with $E > 10^{20}$ eV in extra-galactic space, degradation is rapid. When the thermal nature of the microwave flux is taken into account, the cut-off is quite sharp above about 5×10^{19} eV (Fig. 4.5).

Pair production for 7×10^{-4} eV photons has a proton threshold energy of 7×10^{17} eV. The energy loss by this process is less than by pion production but should cause a slight steepening of the spectrum above 10^{18} eV. Although observations are sparse at these high energies, the results so far do not agree with these predictions. The cosmic-ray spectrum seems to flatten above 10^{18} eV and already one 10^{20} eV primary particle has been detected. Better statistics will be required before any claims on the extent of the 3° field can be made for these results.

The effect of the microwave field has been discussed in more detail by Hillas (1967) who has considered evolutionary effects. If the highest energy cosmic rays are of extra-galactic origin, then the interaction between them and the microwave flux would have been greater in the past when the cosmic-ray sources (radio galaxies) would have been

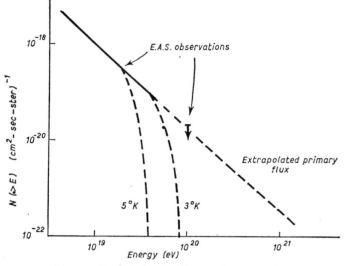

Fig. 4.5. Predicted cut-off in cosmic-ray spectrum

stronger and the black-body temperature would have been greater. If the cosmic-ray production spectrum goes as $E^{-1.5}$, then these effects could lead to a steepening of the spectrum to $E^{-2.2}$ between 10^{16} and 10^{18} eV. The transition at 10^{18} eV could probably be explained by a refinement of this treatment. The cut-off would be about the same as that predicted in Greisen's treatment.

5
Radio Galaxies

5.1 Early history

The development of radio techniques during World War II enabled radio astronomy to make rapid progress in the early post-war years. The first discrete radio source was discovered in 1946 by Hey, Parsons and Phillips in Great Britain; because of its location in the constellation Cygnus, it was called Cygnus A. Shortly afterwards a number of discrete sources were discovered; these included Virgo A, Taurus A, Hercules A, Centaurus A and Cassiopeia A. By 1950 Taurus A had been identified with the Crab Nebula, and Virgo A and Centaurus A with peculiar galaxies. These identifications were possible because of the prominence of a peculiar optical object within the large position error of the radio source; for the other radio sources there were no optical objects immediately evident. At this stage the angular diameter of the position error of a radio source was typically of the order of 10°; within a circle of this angular diameter lay many millions of visible astronomical objects, both stars and galaxies. The ambiguity in relating the radio sources to optical objects could only be reduced by a more accurate location of the radio source.

Radio source locations are limited by diffraction: the uncertainty in position is $\theta > \lambda/d$, where λ = wavelength at which the observation is made, and d = aperture of the radio telescope. Obviously the optimum positional accuracy can be achieved using large telescopes at short wavelengths. In practice, using sophisticated radio techniques, θ can be reduced to seconds of arc. The angular resolving power of an optical telescope is given by $\theta > 1\cdot22\,\lambda/d$ where d is the lens or reflector aperture. Because of the shorter wavelength this limit is unimportant in the optical region; the practical limit is determined by atmospheric scintillations and is the order of a fraction of a second of arc.

In 1951 Smith at Cambridge was able to fix the position of Cygnus A with an accuracy sufficient for Baade and Minkowski to make an optical identification using photographs of this region taken with the 200-inch telescope at Mount Palomar. The optical object had a peculiar appearance, two galaxies apparently overlapping. This led to the suggestion that the Cygnus A radio source might be caused by the collision of two galaxies. Subsequent identification of other extra-galactic radio sources with similar objects led to the early hypothesis that collision was fundamental to radio emission on this scale. This hypothesis has lately fallen into disfavour.

The most significant aspect of these radio sources is the very large energies involved. Cygnus A is situated 3×10^8 light-years from the Galaxy, yet it is one of the brightest objects in the radio sky. The apparent optical magnitude of the double galaxy is only $+15$, 10^{17} times less bright than the sun. Cygnus A is thus a new class of object in which the radio emission is far more significant and energetic than its optical emission; objects of this type are called radio galaxies. Although the number of such sources now stands at hundreds, the nature of the objects is still a mystery. In this respect galactic studies are in the same position as stellar studies prior to the identification of thermonuclear reactions as the principal source of energy release.

5.2 Radio techniques

The central problem in the study of radio galaxies is their identification with an optical object. The spectral lines associated with the optical object can be used to fix the distance to the object if Hubble's Law is assumed. The nature of the optical object is usually peculiar and is an important clue to the workings of the radio galaxy.

The necessity of fixing the radio position as accurately as possible is the major technical difficulty associated with the study of radio galaxies. Since most of the radio galaxies have small angular size and low apparent radio power, high angular resolution must be achieved without the loss of sensitivity. The angular resolution of a pencil-beam antenna, e.g. a parabolic reflector with an antenna at its focus, is set by diffraction; for $\lambda = 1$ m and $d = 100$ m, θ is greater than $0.5°$. In practice the largest steerable radio telescopes have $d < 75$ m. At shorter wavelengths a resolution of $\sim 10'$ can be achieved but more stringent demands are made on the surface accuracy of the paraboloid. Although bigger telescopes are possible and are at the planning stage, the engineering and financial requirements increase rapidly with d. The resolution gained in this way will still be the order of a minute of arc compared with tenths of seconds of arc with optical telescopes.

Hewish (1965) has estimated that the present demands of radio astronomers can only be met by a steerable paraboloid with an aperture of 1·5 km and surface suitable for working at wavelengths down to 20 cm. Even if such a telescope were feasible it would only be a temporary solution. In this respect the position is similar to that in nuclear physics, where the energy limit of accelerators is never sufficient for the nuclear physicists. Fortunately there are ways of achieving high angular resolution without having to go to very large telescopes; some of the ingenious techniques that have been developed will be briefly reviewed.

(a) LUNAR OCCULTATION

The moon in its passage through the heavens behaves essentially like an opaque disc at radio wavelengths, blocking the radio emission from those sources which happen to lie behind it. By a careful study of the variation in intensity of the radio flux as the moon passes in front of the source, the position and size can be determined. As the source reappears further study is possible. The disadvantages of the technique are: (i) there is an inherent ambiguity, since the change in intensity at any instant can be due to occultation at any part of the moon's edge; (ii) only sources in a limited region of the sky can be studied; (iii) occultation of a particular source is a rare occurrence (about once a year) while the duration of the occultation is very short (of the order of minutes); (iv) at short wavelengths the moon itself is brighter than many of the weaker sources.

Despite these disadvantages the lunar occultation technique has made important contributions to radio source studies; in particular the determination of the double nature of 3C 273 in 1962. Since no special equipment is used, the technique has been used at observatories all over the world. Its ultimate limitation is the uncertainty of the topography of the moon's edge which limits the angular resolution to one second of arc.

(b) RADIO INTERFEROMETER

The interference properties of wave phenomena were first used in astronomy in 1923 when Michelson used an optical interferometer to measure the angular diameter of the bright star, Betelgeuse. The essential features of the stellar interferometer are shown in Fig. 5.1; the mirror used was the 100-inch reflector at Mount Palomar. The angular resolution is given by $\theta = \lambda/D$ where D = separation of the mirrors. In theory θ can be made as small as required by increasing D;

F

in practice mechanical considerations limit D to a few metres, since the light paths must be equal to within a few wavelengths. It has not been possible to make $\theta <$ o·oı″. Since many stars of interest have angular

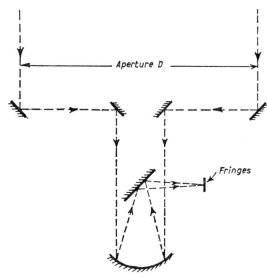

Fig. 5.1. Michelson interferometer

diameters less than this, the stellar interferometer has had only limited application in optical astronomy.

Radio interferometers, using essentially the same principle as the Michelson stellar interferometer, have made a relatively much more significant contribution to radio astronomy because of the severe diffraction limitation of pencil-beam antennas. Radio interferometers were first used in 1946; the essential features are shown in Fig. 5.2.

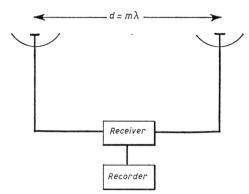

Fig. 5.2. Radio interferometer

Signals from the two antennae are taken to a central station. The interference pattern is observed in the receiver output as the source transits due to the difference in path of the plane wave from the radio source to the two antennae. The receiver output represents the Fourier transform of the radio intensity distribution across the source in a direction parallel to the line joining the two antennae. The angular resolution is $\theta \sim 1/m$ where the separation between the aerials is $d = m\lambda$. Not only is the angular resolution greatly increased compared with a single antenna, but the sensitivity of the system to a point source is significantly greater.

The long base-line radio interferometer, used by the Jodrell Bank radio astronomy group, is an important example of this technique. The two elements consist of the 250-ft parabolic reflector at Jodrell Bank and the 84-ft parabolic reflector at the Royal Radar Establishment, Malvern, 79 miles (127 km) away. At a wavelength of 0·21 m, $m = 600,000$. The central station is at Jodrell Bank, with signals from the Malvern station coming by microwave link via two repeater stations. This instrument was developed primarily to resolve a small number of quasi-stellar objects, too small to be resolved by an earlier instrument. Even at this high resolving power ($\sim$0·1″) at least five of the radio sources are unresolved.

The latest development in this technique is the use of video tape recorders with standard clocks at stations separated by over a thousand miles. The tapes are then analysed by computer and angular resolutions of 0·01 seconds of arc are possible.

(C) APERTURE SYNTHESIS

The same principle is utilized in the aperture synthesis technique developed at Cambridge by Ryle. This consists of two or more antennae, which may be moved relative to one another. With one antenna fixed, the others are moved through successive positions until observations have been made with the elements at all the positions that would be covered by a very large antenna of total area equal to that swept out by the elements. As in the simple radio interferometer the individual antennae can be relatively small and great flexibility is possible.

The chief disadvantage of the technique is the long observing time necessary since observations must be repeated with the antenna in a variety of positions. However, the scanning rate of a pencil-beam antenna is also slow since observations of a particular point must last at least as long as the integration time of the receiver. These techniques are not suitable for the study of variable or short-lived phenomena.

The most sophisticated aperture synthesis system is at Cambridge. It consists of three steerable 60-ft parabolic reflectors, two of which are fixed at a separation of 2500 feet and the third movable on a 2500-ft track (Fig. 5.3). The output from the movable antenna is simultaneously combined with that from the fixed elements, thus reducing the total observing time. Although the three antennae lie on an East–West axis, observations made over a twelve-hour period enable an elliptical

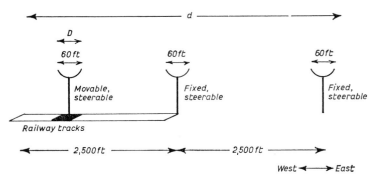

Fig. 5.3. Cambridge aperture synthesis

ring of a large equivalent aerial to be synthesized due to the rotation of the earth. A map of an area of one square degree requires two months of observation; at a wavelength of 21 cm an angular resolution of 23 seconds of arc is obtainable. It is estimated that the sensitivity is the same as that obtained with a conventional aerial of area $3Dd$ where D = diameter of the interferometer elements and d = maximum separation of the elements.

5.3 Radio source surveys

Unlike the optical survey of the sky where a large area can be seen with one exposure of a photographic plate, a radio survey is a much slower and more complicated process. If the angular resolution of the instrument is θ, observations must be made in directions separated by no more than $\theta/2$. Each observation must last at least as long as the integration time of the instrument, that is, the time required to detect a signal of a particular sensitivity above noise. From these observations a map of the radio sky can be drawn in which the radio source positions are fixed to within θ degrees. Unfortunately these surveys are complicated by a number of effects:

(i) The interferometers usually used are deliberately designed to give the maximum response to sources with small angular diameters.

Sources with diameters greater than θ are discriminated against and may be omitted altogether.

(ii) An inevitable feature of any radio telescope is the presence of side lobes; the presence of a strong source in one of these can be interpreted as a weak source in the main lobe. This is particularly important in the vicinity of the brightest radio sources.

(iii) Due to the relatively poor angular resolution there can be confusion when several weak sources lying close together are interpreted as one stronger source.

(iv) Since the surveys are normally made at one frequency with a narrow bandwidth, surveys of the same region of the sky at different frequencies will not show the same sources due to variations in the source frequency spectra.

The net result of these effects is that the positions or existence of a large number of radio sources has been disputed. The second catalogue of radio sources compiled at Cambridge in 1953 (the 2C catalogue) listed 2000 sources; later observations showed that 75% of these were erroneous. Only where surveys overlap can the results be treated with confidence. Fortunately, the above effects are now better understood so that most of the source positions are now regarded as established.

Most radio sources in the northern hemisphere are designated by their number in the Third Cambridge Catalogue of radio sources, which was completed in 1959. The limiting sensitivity was 9 f.u. (1 f.u. $= 10^{-26}$ watts/m^2/c/s) and the positional accuracy for over 300 sources was better than a degree. 90% of these sources are believed to be extra-galactic. This survey is now being replaced by the Fourth Cambridge Catalogue. Using the aperture synthesis method the limiting sensitivity is 2 f.u.; a small region of the sky has been surveyed down to 0·25 f.u. Over 5000 sources have been located with a positional accuracy of minutes of arc. Other catalogues in the northern hemisphere include the CTA, NRAO and Bologna.

In the southern hemisphere the most complete catalogue is that made using the Parkes 210-ft parabolic reflector in Australia. Over 2000 sources are listed with positional accuracy of ± 1 minute of arc. The limiting sensitivity is 2·5 f.u. Observations were made at three frequencies (408, 1410, 2650 Mc/s) so that the reliability of the catalogue is high.

Sandage (1966) estimates that the recent surveys indicate that there are 10,000 sources over the universe with flux densities at the earth greater than 1 f.u.

5.4 Classification of extra-galactic radio sources

The optically identified radio sources are most meaningfully classified in terms of their optical forms. Matthews, Morgan and Schmidt (1964) have given a thorough discussion of 52 optically identified extra-galactic objects which were established prior to 1964. While this study now requires some revision to include the subsequent identifications (the number is rapidly growing and already surpasses two hundred) their inclusion does not materially affect the results of this analysis. The degree to which an identification can be considered certain depends somewhat on a subjective judgement; these authors adopt more rigorous criteria than usual and claim that the 52 sources listed are definitely established. Their three criteria are:

(*i*) *Position.* The radio and optical centroids must be reasonably close. Very often the radio source is double; in that case the optical counterpart should lie halfway between the two radio objects.

(*ii*) *Appearance.* The optical source should display some unusual features. Experience has shown that there are certain characteristics usually associated with the optical object; these include double galaxies, ejected wisps of gas, ultraviolet excess, large luminosity compared with other objects in a cluster, star-like appearance.

(*iii*) *Optical spectra.* Almost all extra-galactic radio sources display strong emission lines, often considerably broadened, in their optical counterparts.

In some cases one of these criteria may be relaxed if the other two criteria are completely satisfied. In general the identifications were based on radio source positions established in one of the radio surveys mentioned above. The optical objects were usually located on the plates of the Palomar Sky Survey; this covered the whole sky north of a declination of $-33°$ with 940 red and blue plates taken with the 48-inch Schmidt camera. In some cases special plates were taken of the area of interest with the 200-inch telescope at Mount Palomar.

For each of the 52 sources the following information is given:

(i) The distance d based on red-shift measurements (in some cases this is not available and an estimate has been made from the visual magnitude).

(ii) The radio luminosity L defined as

$$L = 4\pi d^2 \int_{\nu_1}^{\nu_2} I(\nu)\, \mathrm{d}\nu$$

where ν = frequency and $I(\nu)$ = flux density = $k\nu^{-\alpha}$ where k and α are constants for the source. k includes a cosmological correction depending on z. ν_1 and ν_2 are taken as 10^7 and 10^{11} c/s respectively.

(iii) The optical form of which seven varieties are considered: (*a*) spirals and irregulars, (*b*) ellipticals, (*c*) D galaxies (elliptical nucleus surrounded by visible envelope), (*d*) dumbbell (two approximately equal nuclei within a common envelope), (*e*) N galaxies (very bright nucleus with faint envelope), (*f*) quasi-stellar sources (Chapter 6), (*g*) E + D galaxies (intermediate between (*b*) and (*c*) above).

The distribution of radio luminosities in these groups is illustrated in Fig. 5.4. Part of this distribution is influenced by selection effects; in

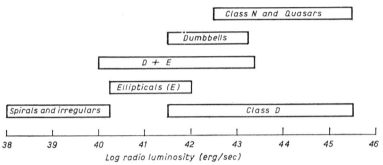

Fig. 5.4. Radio galaxy types as a function of luminosity

particular large, but weak, sources will not be picked up in the radio surveys in which interferometric or aperture synthesis techniques were used. There are probably more sources in the $L = 10^{41}$ ergs region than are included in this list.

Two broad divisions are apparent:

(i) Weak sources with $L \leqslant 10^{40}$ ergs. All the sources in this group are class (*a*). These are essentially weak radio emitters and are comparatively close. The radio source is invariably single and close to the optical object. The Galaxy falls into this class. NGC 1068 is included although it has many of the characteristics of group (ii).

(ii) Strong sources with $L > 10^{40}$ ergs. The sources in this group are usually called radio galaxies. Most of the sources are double; however, the dumbbell galaxies are always associated with a single radio source. The characteristics of the most prominent members of this group are considered in 5.10.

A large number of the radio galaxies occur in clusters in which they are usually associated with the most prominent member; other clusters are observed with almost identical optical characteristics but with no detectable radio emission.

5.5 Radio spectrum

After the location of a source, the next step is to measure the flux density $I(\nu)$ (measured in watts/m²/c/s) as a function of frequency. Since the atmosphere is transparent to radio waves over many orders of magnitude of frequency, these measurements involve a large range of radio techniques, with aerials ranging from low-frequency dipole arrays to parabolic reflectors and eventually to giant horn antennae at microwave frequencies. At all frequencies measurements are difficult to make absolutely. It is comparatively easy to make measurements relative to some other source which can be regarded as a standard. The flux density spectrum is usually an amalgam of data obtained at various frequencies in different observatories. The absence of absolute standards makes the combination of these data difficult. In these circumstances it is surprising that radio spectra of a large number of sources have been established reliably. This circumstance is due largely to the power law

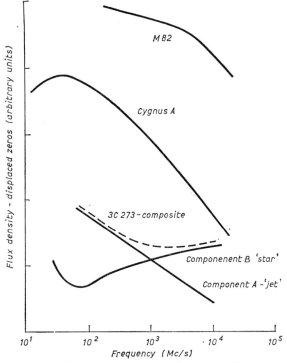

Fig. 5.5. Radio spectra of three well-known radio sources

nature of most radio source spectra; the spectral index, α, is usually constant over the complete radio range. Hence if observations are made at only a few widely scattered frequencies, it is sufficient to interpolate between these points.

1. The quasi-stellar radio source 3C 48. At first sight the image is identical to that of a star of 16th magnitude. Only the large red-shift $(z = 0.37)$ and strong radio emission reveal its unusual nature. On closer examination a faint red nebulosity is observable around the star-like object.

(*Mount Wilson and Palomar Observatories*)

2. The quasi-stellar radio source 3C 273. One of the first quasars discovered, it is also one of the closest ($z = 0.16$). Note the faint jet (component A) extending towards the bottom right hand corner; this is the strongest component at radio wavelengths.

(*Mount Wilson and Palomar Observatories*)

3. One of the 6 m² liquid scintillators of the vast Sydney array which, when completed, will have a collection area of 250 km². With this array it is hoped to investigate the upper energy region of the cosmic ray spectrum.

(Courtesy of C. B. A. McCusker)

4. Each element of the Sydney array is composed of two large scintillators, separated by 50 m and operated in coincidence; outputs are stored on magnetic tape together with a timing signal from a central transmitter. The tapes will be collected weekly and analysed by computer. The spacing of the elements in the array lattice varies from 0·4 to 1·6 km. It is located in the Pilliga State Forest, near Narrabri, New South Wales.

(*Courtesy of C. B. A. McCusker*)

Originally all flux measurements were made relative to Cassiopeia A, whose flux was measured absolutely at a number of frequencies. By interpolation this source can be used as a standard at all frequencies. Unfortunately the intensity of Cassiopeia A is variable with time; it decreases by 1% per annum. It has now been replaced as a standard by a number of weaker sources.

Most radio galaxy source spectra can be represented by a power law between 178 and 1400 Mc/s. The average value of the spectral index, α, is 0·9 but values vary from 0·4 to 1·9 (see Chapter 6.2). Below 178 Mc/s many sources exhibit a sharp fall-off of flux with decreasing frequency; this indicates that synchrotron self-absorption is taking place within the source. This effect is well understood and is expected for high surface brightness. Above 1400 Mc/s many of the sources have a significant change in α. In some cases distinct maxima and minima can be seen in the radio spectra (Fig. 5.5).

5.6 Radio brightness distribution

As the angular resolution of radio telescopes improves, radio source structure can be examined. The degree of resolution is now so good that only in a few cases are sources unresolved; in the other sources some sort of structure can be seen. In a few cases the source has a Gaussian structure; somewhat more common are those with bright central regions surrounded by a halo. The most common structure is radio doubling, that is, two similar regions separated by a distance greater than their diameter. This doubling feature is found in about 70% of the identified sources and is apparently fundamental to radio-galaxy evolution. In general the optical object identified with the radio galaxy lies near the centre of the line joining the two radio regions. Typical dimensions of these regions are 25 kpc with a separation of 100 kpc. The optical object does not enclose the radio sources, having a diameter of about 25 kpc (Fig. 5.6). This structure suggests that the two radio regions are the remnants of an explosive event which caused them to be ejected from the optical galaxy in opposite directions. These explosions may be recurrent.

A number of sources have been observed with a jet extending from the central object; this also seems to be evidence in favour of an explosive origin.

A typical angular diameter of an extra-galactic radio source is one minute of arc; a small number of sources have diameters less than one-tenth of a second of arc. For the closer sources it is now possible to obtain detailed maps of the radio distribution. Centaurus A, for

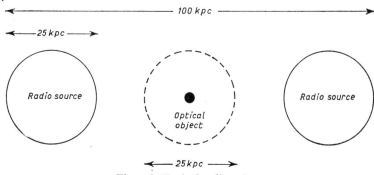

Fig. 5.6. Typical radio galaxy

example, has been shown to consist of two components separated by 10°; in addition there are two smaller radio regions almost coincident with the optical object. As radio techniques improve, all radio sources will probably show complex structures.

5.7 Polarization

Some degree of polarization can be detected from most radio galaxies at frequencies greater than 10^9 c/s. The amount of polarization increases with frequency; in some cases it is as much as 30%. In certain radio galaxies polarization is also detected at optical wavelengths. As mentioned in Chapter 2, polarization is normally interpreted as evidence for the magnetic bremsstrahlung origin of the radiation. Polarization measurements are extremely difficult; in general all measurements must be checked against a similar source known to be unpolarized to isolate instrumental effects.

From a study of the direction and strength of the polarization at various points in the source, some idea of its magnetic structure can be derived. However, in traversing a magnetic field, the angle of the electric field vector undergoes a rotation. This Faraday rotation, both in the outer region of the source and in the Galaxy, complicates the interpretation of these measurements. Since Faraday rotation is proportional to (frequency)$^{-2}$, the true direction of polarization can be determined by making measurements at at least two frequencies. So far these detailed measurements have only been made on the nearest, and hence brightest, radio galaxies.

The results from these measurements have been surprising in that they indicate a significant change of magnetic field structure with the age of the galaxy. In young double radio sources the magnetic field vector is parallel to the line joining the two sources; in older galaxies the vector is perpendicular to this apparent direction of motion. That polarization is detected at all indicates a high degree of regularity, but does not give any measure of its strength which can only be determined by assuming equipartition between the various energy forms (2.3).

These polarization studies are only in the preliminary stage; eventually they may provide important pointers to the evolution of a radio galaxy subsequent to the ejection of the two radio-emitting regions. It is unlikely that much information about the actual explosive event will come from these studies.

5.8 Optical spectra

The optical spectra of radio galaxies are characterized by strong emission lines. This circumstance is fortunate, for without these lines red-shifts would have been difficult to obtain and hence the distance and size of the objects would be unknown. The information obtainable from the spectral lines on the physical conditions within the source have been summarized by Burbidge (1966).

The spectral lines come both from the stars in the source and the interstellar gas. The emission lines come from the gas and it is these that are most important. From these the electron densities, the temperature, ionization conditions and relative abundances of the elements can be deduced. In addition measurements of the Doppler shift at various points in the source indicate the structure and relative motion of various parts of the source.

The spectral lines observed are similar to those from gaseous nebulae around hot stars in the Galaxy. Hydrogen and helium are observed; forbidden lines of oxygen, nitrogen, sulphur and neon are also strong. Although most radio galaxies exhibit some of these lines, there are some which do not. Also there are many galaxies which have spectral lines similar to those associated with radio galaxies but which are not strong radio emitters.

5.9 Total energies

The very large total energies involved in radio galaxies can be arrived at in a number of ways. A lower limit can be obtained from three experimental observations: (i) the apparent radio luminosity (from 10^7 to 10^{11} c/s), (ii) the red-shift assuming Hubble's Law, (iii) the angular size. This simple method, which involves no assumptions about the mechanism of emission, is summarized below.

There are five steps:

(1) Accurate radio position $\longrightarrow$ optical identification and therefore red-shift

(2) Apparent radio luminosity and red-shift $\longrightarrow$ absolute radio luminosity

(3) Angular dimensions and red-shift $\longrightarrow$ linear dimensions

(4) Linear dimension/velocity of light $\longrightarrow$ minimum age

(5) Total energy radiation = absolute luminosity $\times$ age.

The total energy arrived at by this method is an underestimate because (i) only the radiation at radio frequencies has been considered, (ii) the age has been estimated assuming that the source expanded with the velocity of light, (iii) the luminosity is assumed to have been constant since the outburst, whereas it appears more likely in an explosive object that it has been decreasing.

The absolute radio luminosities range from 10^{38} to 10^{45} ergs/second. For a typical lifetime of 10^6 years, the minimum total energies range from 10^{51} to 10^{58} ergs.

A more realistic estimate can be made using the magnetic bremsstrahlung theory of radiation and the equipartition method (2.2). Values obtained by this method for the principal radio galaxies are summarized in Table 5.1.

TABLE 5.1

Extra-galactic radio sources

SOURCE	H (gauss)	U (ergs)
Cygnus A	2×10^{-4}	3×10^{60}
Centaurus A	4×10^{-6}	2×10^{59}
Virgo A	2×10^{-5}	10^{58}
M 82	$2 \cdot 5 \times 10^{-6}$	10^{59}
Perseus A	3×10^{-6}	8×10^{59}

The interpretation of these large energies and a discussion of the possible energy release mechanisms will be deferred until Chapter 7, since the recently discovered quasars are pertinent to any discussion of radio galaxies.

5.10 Well-known radio galaxies

The principal characteristics of some of the most studied radio galaxies are listed below and in Tables 5.1 and 5.2.

(A) CYGNUS A

Two identical radio components, separation 80 kpc; elongated and non-uniform; brightness greatest at the extremities; third brightest

TABLE 5.2

SOURCE	LOG RADIO POWER (watts)	DISTANCE (Mpc)	RADIO SOURCE	OPTICAL SOURCE
Cygnus A	38	170	Double	Double
Centaurus A	35	5	Double	Single
Virgo A	35	11	Jet	Jet
M 82	33	3	Single	Exploding
Perseus A	35	55	Single	Seyfert

source in sky (after sun and Cass A); optical source midway between radio sources; elliptical halo with two small optical objects, apparently colliding; structure similar to that of many other radio galaxies; separation of radio sources from central optical object suggests that they have been ejected.

(B) CENTAURUS A

Radio source: large double source, components elongated; distance from outer edges 770 kpc (about 10°); separation of maxima of these components is 240 kpc; smaller double source near the centre, coincident with optical source. Optical source: large spherical galaxy extending 80 kpc along axis of radio source; thick dust lane crosses galaxy at right angles to major axis; thought at one stage to be collision between elliptical and spiral galaxies; from relative Doppler shifts, source is rotating about major axis of radio sources.

(C) VIRGO A (M 87)

Nucleus of optical galaxy has blue jet (about 1 kpc in length) extending from centre; radio double source coincident with jet and nucleus: superimposed on these is large elliptical E_0 (almost spherical) galaxy, with diameter of about 35 kpc; single radio source coincident with this large optical object; optical continuum in jet strongly polarized; x-ray source in this region which may be coincident with this peculiar galaxy; some features in common with quasar 3C 273.

(D) M 82

Young galaxy, believed to be observed only one million years after explosion; large gas jets observed from above and below principal plane; velocities of gas in jets is about 1000 km/sec; optical continuum strongly polarized; radio power intermediate between weak and strong galaxies; believed to be the most definite evidence for occurrence of explosion on a galactic scale; total energy involved estimated as between 10^{56} and 10^{59} ergs; energies of radiating electrons at least 10^{11} eV which suggests that the object is significant source of cosmic radiation.

(E) PERSEUS A (NGC 1275)

Radio source: complex; radio spectrum has distinct minimum; may be composite of two sources; radio source smaller than optical source; occurs in cluster of galaxies; two other radio sources close by, which may be associated. Optical source: Seyfert galaxy (bright nucleus surrounded by broad region of strong optical emission); once believed to be collision of two galaxies; gas velocities of about 3000 km/sec; evidence for recent explosion similar to that in radio galaxy; may be at different stage than normal radio galaxy since there are some unusual features; only one other Seyfert galaxy is a radio source.

6
Quasars

6.1 Discovery

The discovery of quasi-stellar radio sources or quasars, as they are now generally called, ranks as one of the most exciting stories of scientific investigation in this decade. Whatever the origin of these strange objects their early history has provided many surprises and stimulated a concentration of effort never before seen in astronomy. The enigmatic nature of the quasars was revealed after a series of experiments, in which optical and radio astronomers of many nationalities co-operated. The great interest that was generated as a result of their discovery has given added incentive to all astronomical studies. Whether the ultimate explanation of the phenomena will cause a fundamental revision of current astronomical, or perhaps even physical concepts, the very great effort that has been expended in their study makes their discovery one of the important milestones in the history of astronomy.

The extraordinary nature of these objects was not apparent from the preliminary radio surveys in the 1950's which catalogued many of the quasars; it was only after the size of some of these sources was investigated, that they appeared as a distinct phenomenon. A survey of radio sources in the 3C catalogue was made in 1960 using the long baseline interferometer of Jodrell Bank. The average size of the sources examined was 30 seconds of arc. The resolution of the interferometer at this stage was one second of arc; a small group of sources were found to have angular diameters less than this. With the position defined to this accuracy, optical identification was comparatively straightforward. Four of these objects, 3C 48, 3C 147, 3C 196, 3C 286, were tentatively identified with optical objects which appeared stellar, rather than galactic, in scale.

The existence of radio stars had been proposed in the early stages in the development of radio astronomy. The proposal fell into disfavour when it was realized just how weak the radio emission from the sun was in absolute terms; with the identification of several radio sources with galaxies, it was assumed that radio stars were non-existent. Short-lived phenomena, such as flare stars or supernovae, were exceptions, but in these, the optical object was obviously not a quiescent star. The identification of these four objects with star-like objects immediately reopened the radio-star debate; it was not clear just what kind of star would give this strong radio emission. The attention of optical astronomers was now directed to these 'stars'; using the most sophisticated optical telescopes at Mount Palomar, the spectra and structure of the optical sources were examined. Sandage (1961) found an extensive, but faint, nebulosity surrounding the star-like object associated with 3C 48; its spectra showed several diffuse emission lines superimposed on a continuum. The emission lines did not coincide with those of any known element. It was obvious that these were no ordinary stars.

At the same time that these optical investigations were being pursued, a group of Australian radio astronomers used the lunar occultation technique to determine the position and structure of 3C 273 with unprecedented accuracy [Hazard (1963)]. 3C 273 was one of the sources isolated by the Jodrell Bank survey but had not yet been optically identified. The radio source was shown to be double; it was optically identified with a star-like object, which was coincident with one of the radio components (B) (Fig. 6.1). The other component (A) was coincident with a jet-like nebulosity. The star-like object had a magnitude

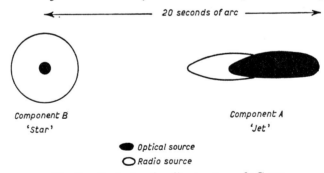

Fig. 6.1. Optical and radio structure of 3C 273

of +13 which was brighter than any of the other quasi-stellar objects. Component A is the stronger of the two radio components but is barely visible optically. It is now believed that B is the parent object, A having been emitted some time ago.

The importance of this identification lay in the exact coincidence of the double optical and radio sources which removed any ambiguity in the identification. The previous identifications had been only tentative;

since component B was similar in every way to these objects, all the identifications could be regarded as certain. Schmidt (1963) examined the spectra of these objects; for each of the five objects weak emission lines could be seen. However, no two sources had even one line in common. Of the six lines visible in the spectrum of 3C 273, four formed a harmonic pattern suggesting that they came from an atom with just one electron. No known element had the same spectrum. Schmidt made the surprising discovery that if the lines were shifted to the red by 16%, then the four lines could be identified with the spectrum of hydrogen. A similar shift enabled the other two lines to be associated with O III and Mg II.

In the light of this discovery Greenstein and Matthews (1963) re-examined the spectra of 3C 48. If a red-shift $z = 0.37$ was assumed, then six of the lines could be identified. These were Mg II and forbidden lines of O II, Ne VI, and Ne V(3). Unlike 3C 273 there were no strong hydrogen lines indicating that there were important differences between the two objects. With the discovery of these red-shifts the enigma of the quasar phenomenon became apparent; although stellar in appearance, the objects had red-shifts which, if interpreted in accordance with Hubble's Law, would place them on the edge of the observable universe. Large red-shifts had previously been observed in a number of radio galaxies, e.g. Cygnus A, $z = 0.057$; 3C 295, $z = 0.46$. But the angular size of these objects had confirmed their galactic scale, and hence their high luminosities without which they would have been undetectable. Assuming Hubble's Law, 3C 273 and 3C 48 are at distances of 470 and 1100 Mpc respectively. To be detectable at these large distances their absolute luminosities must be similar to that of radio galaxies. However, if 3C 273 were a galaxy, it would be evident on photographs taken with the largest telescopes. In fact it is too small to be resolved by the largest optical telescope. It thus became apparent that quasi-stellar objects represented the release of large energies in relatively small volumes. Just how small these volumes are was demonstrated by the variability of the optical luminosity of these objects. Examination of photographic plates taken in sky surveys over the last seventy years showed that the brightness of 3C 273 has varied by as much as 50% over a period of a few years. These observations set a definite upper limit to the size of the emitting regions. It is not possible to observe coherent changes with a time-scale shorter than the dimensions of the region of emission in light-years, i.e. the change cannot be propagated through the source with a velocity greater than the velocity of light. The optical variations in 3C 273 therefore indicated that the diameter of the quasar was of the order of 1–10 light-years.

Galaxies have diameters of the order of tens of thousands of light-years. Making the usual assumption of a magnetic bremsstrahlung origin of the observed continuum, the total energies involved in quasi-stellar sources can be estimated for an equipartition model. These energies are of the order of 10^{60} to 10^{62} ergs. The release of this amount of energy corresponds to the complete annihilation of 10^6 to 10^8 stars with masses similar to that of the sun.

Thus, although the volumes associated with quasars may be only 10^{-9} times those of a normal galaxy, their radio luminosities are as much as that of the largest radio galaxies. Five years have elapsed since this phenomenon was revealed, but the intervening years have added little to the elucidation of its nature. On the contrary the subsequent discovery of some hundreds of quasars and the investigation of their properties has increased the complexity of the problem. There have been many surprises and it is clear that quasars represent a major constituent of the observable universe. The principal observational data will be summarized in the following sections; the various theories that have been suggested will be discussed in Chapter 7.

6.2 Radio spectrum

The radio spectrum for quasars is very similar to that for radio galaxies, closely following a power law with exponent α. Kellermann has considered the distribution in the spectral index α for several hundred extra-galactic radio sources – both radio galaxies and quasars. The mean value for α over the range 38 to 1400 Mc/s was taken; α varies from 0·2 to 1·3 with a mean of 0·77. This large spread in α suggests that it may be possible to correlate the value of α with some other property of the sources. There is no evidence for any dependence on the apparent luminosity; however, a plot of α against the linear dimensions of the sources (assuming the cosmological red-shift) shows that α increases with the source dimensions. The source size may, in turn, depend on age, indicating that the spectrum steepens with age. It has been suggested that the absolute luminosity of radio galaxies increases with α; quasars do not show this dependence, having a small range of absolute luminosity for a large spread in α.

There is some evidence for a radio absorption feature (0·5%) in the continuum of 3C 273. This broad feature is the well-known 21 cm line of hydrogen; the broadening corresponds to a Doppler shift in a hydrogen cloud which has a velocity of the order of 1400 km/sec. This cloud of gas may be part of the Virgo cluster which lies in the same area of the sky as 3C 273. Since the distance to the Virgo cluster is well-established as 12 Mpc, this absorption feature, if real, sets a definite lower limit to the distance to 3C 273, independent of the cosmological interpretation of the red-shift.

6.3 Radio size and structure

The long base-line interferometer at Jodrell Bank has been used to measure the size and structure of a number of radio sources [Adgie *et al.* (1965)]. Observations have been made at wavelengths of 21, 11 and 6 cm. Fifty sources, which previous surveys had indicated had small angular sizes, were studied; most of these were quasars. The angular resolution of the instrument was 0·025 seconds of arc; even at this high resolution most of the sources were unresolved. The maximum dimensions indicated by these studies assuming the cosmological red-shift are of the order of 10–100 parsecs. These small dimensions are in agreement with the radio variations (6.4). All of the sources which show variations with short time constants are unresolved. In particular the B component of 3C 273, one of the closest quasars, is unresolved at all frequencies indicating a high brightness temperature.

6.4 Radio variations

The variability of the radio and optical emission from quasars has been one of their most surprising features. The existence of these variations has caused serious doubts about the position of the quasars; if they are at the distances indicated by their red-shifts then the sizes indicated by the time-scale of the variations seem prohibitively small.

Radio variations are particularly difficult to establish because of the absence of absolute standards, the narrow bandwidth of the observations and the use of a variety of techniques at observatories scattered over the globe. Until recently most variations were put down to instrumental effects. The first radio variation of a quasar was announced by a Russian group in 1965; variations in CTA 102 with a period of 100 days were reported at a frequency of 1000 Mc/s [Sholomitsky (1965)]. It is a measure of the uncertainty pertaining to variation measurements that this result has not been confirmed by other groups and is now discredited.

Since then, reliable and confirmed variations have been observed in a small number of quasars at high frequencies. The N.R.A.O. group at Greenbank, West Virginia, have observed 78 radio sources of small angular diameter at wavelengths of 22 and 40 cm in 1963 and 1966. Nine sources, four of them quasars, have shown significant variations over this period. These results, combined with those at higher frequencies, indicate significant changes in these sources with periods of the order of ten years. One of the most interesting of these sources is 3C 273. At a wavelength of 40 cm, the emission has remained approximately constant between 1962 and 1966; over the same period the emission at shorter wavelengths had undergone a considerable increase. At 2 cm a hundred per cent increase has been reported between 1963 and 1966. In 1966 3C 273 was observed at 2 cm over a five-week period; the flux density increased by $1 \cdot 7 \pm 0 \cdot 5$ f.u. per week. The Seyfert Galaxy, NGC 1275, underwent variations similar to 3C 273 over the same period. The emission of 3C 279, a quasar, at 2 cm passed through a minimum in the period 1964–5 and is now increasing.

These large variations over short periods are indications of violent events in concentrated regions. Pauliny-Toth and Kellermann (1966) propose that the complex spectra of these sources result from the injection of bursts of relativistic particles ranging from days to years. Similar injections may take place in radio galaxies but at less frequent intervals. These authors point out that energies involved are very great; the spectrum of 3C 279 requires the injection of 3×10^{58} ergs in the form of relativistic electrons in less than ten years. The energy output is equivalent to that of 10^8 Type I supernovae.

The study of radio variations is still in its infancy. Since the demands on the observing time of the large radio telescopes are severe and since observations of variations to be meaningful must be made regularly at as many frequencies as possible, a full study of radio variations is beyond the capabilities of one observatory. It is unlikely that the full scope of these studies will be realized until a regular observing programme is undertaken on an international scale over a period of years. If the results obtained so far are indicative of the fruits of such an enterprise, the effort will be worth while.

6.5 Optical continuum

The most striking characteristic of the optical continuum from quasars is the strong emission in the ultraviolet. This property has enabled the rapid optical identification of quasars once the radio position has been determined. Experimentally the optical continuum has been studied by taking three sets of plates, with filters which define the ultraviolet (U), blue (B) and visible (V) intensities centred at 3500, 4500 and 5500 Å respectively. From these plates a two-colour index diagram can be drawn in which U–B is plotted against B–V (Fig. 6.2). On this diagram an object with a black-body spectrum lies along a straight line. In general the stars do not lie along this line. Certain classes of stars have a black-body spectrum, e.g., white dwarfs. The quasars also lie close to the black-body line, indicating an excess in the ultraviolet compared

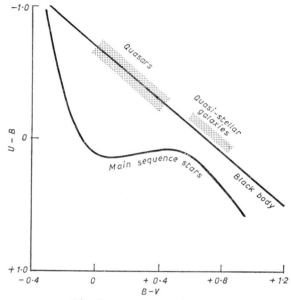

Fig. 6.2. U–B v B–V diagram

to stars of comparable apparent magnitude. This property has led to the development of a convenient method of optically identifying quasars when the radio position is sufficiently ambiguous to include a large number of star-like objects. Plates are taken of the region of interest, one using an ultraviolet filter, one with a blue filter. The exposure times are adjusted such that the images of normal stars will appear approximately equally bright on both plates. Quasars will be immediately obvious using a comparison technique because of their relatively stronger image on the ultraviolet plate.

Using this technique Sandage found a number of objects on photographic plates which resembled quasars in every respect except that they were not detectable at radio wavelengths. These objects were called interlopers or quasi-stellar galaxies; they are now usually classified with quasars under the general heading, quasi-stellar objects. The orginal estimate of the number of these objects, 500 for every quasar, has now been reduced to 80 to 1. They may represent a later stage of quasar development in which the radio emission is no longer detectable.

One particularly interesting result has come from the study of the continuum of one of the most distant quasars. 3C 9 has a red-shift of

2·012 so that the strong Lyman-α line of hydrogen, which is normally observed at 1216 Å, is shifted to 3663 Å. The inter-galactic neutral hydrogen density is estimated, by indirect methods, to be about 10^{-29} g/cm³. The passage of continuum radiation through this hydrogen will give an absorption line at 1216 Å. In the case of distant objects like 3C 9 sufficient hydrogen should be traversed to give a detectable effect. Since the radiation from this distant quasar is red-shifted, the absorption will be evident not as a line but as a general deficiency of continuum radiation at wavelengths between 1216 and 3663 Å. The absorption takes place at 1216 Å but the photons absorbed would arrive at the earth at a wavelength shifted by an amount appropriate to the distance between the point of absorption and the earth. This effect was predicted by Scheuer (1965), and Gunn and Peterson (1965). Experimental measurements have shown no absorption. If the red-shift of 3C 9 has a cosmological origin, then the lack of absorption indicates an upper limit to the neutral hydrogen density of 10^{-35} g/cm³. This value is considerably lower than expected and may be evidence for an alternative explanation of the quasar red-shift. The missing density of hydrogen could, however, be made up by large concentrations of ionized hydrogen.

6.6 Optical spectra

Although quasars are comparatively weak optical objects, their optical spectra have been intensively studied. The most prominent feature is weak broad emission lines superimposed on the continuum with its ultraviolet excess. Recently some absorption features have been detected.

From an analysis of the identity of the emission lines and their relative intensities some idea of the relative abundances of the elements and the conditions in the emitting regions can be ascertained. The abundances indicated are in fair agreement with the cosmic abundance. Emission lines from the lighter elements up to sulphur are detected with the usual absence of Be, B and Li. Electron temperatures are between 10^4 and 10^5 °K and densities are from 10^4 to 10^7 particles/cm³.

The spectra of the two most studied quasars, 3C 48 and 3C 273, have been considered in detail by Greenstein and Schmidt in a classic paper in 1964. The identified lines from the stellar object (component B) in 3C 273 are listed in Table 6.1; several broad bands are also observed which may correspond to groups of lines which are superimposed. O III is the only forbidden line. Within the width of the emission lines (50 Å) all exhibit the same red-shift.

TABLE 6.1
Emission lines from 3C 273

OBSERVED WAVELENGTH	WAVELENGTH AT EMISSION	IDENTIFICA-TION
Å	Å	
3239	2798	Mg II
4595	3970	H$_\varepsilon$
4753	4101	H$_\delta$
5032	4340	H$_\gamma$
5632	4861	H$_\beta$
5792	5007	[O III]

3C 48 is a star-like object with an irregular distribution of faint red wisps surrounding it. The continuum extends well into the ultraviolet and many emission features are evident. Most of the lines are forbidden. Two hydrogen lines are just detectable, unlike the spectra of 3C 273 in which the hydrogen lines are the most prominent spectral feature. The width of the lines suggests velocities of about 1000 km/sec.

The detection of forbidden lines is an indication of low electron densities. They arise from transitions from excited states of atoms with long lifetimes (of the order of seconds). In the laboratory discharge tube, where the density is more than 10^{15} particles/cm³, these states are de-excited by collisions so that lines characteristic of these states are not observed, that is, the lines are forbidden. In astronomical objects the densities are much lower and the lines can be observed. The values of electron density deduced by Greenstein and Schmidt for 3C 48 and 3C 273 are shown in Table 6.2. Also shown are the predicted radius and mass of the emission regions.

TABLE 6.2
Parameters of quasi-stellar radio sources 3C 48 and 3C 273

Source	3C 48	3C 273
Visual Luminosity (ergs/sec)	10^{45}	4×10^{45}
Radius (optical and radio) (parsec)	<2500	<500
Electron density (cm⁻³)	<3 × 10⁴	3 × 10⁶
Mass ($M/M_\odot$)	>5 × 10⁶	6 × 10⁵

The minimum total mass of the quasar can be estimated from the width of the spectral lines. These Doppler widths correspond to gas velocities of about 1500 km/sec. For gas molecules with this velocity to be gravitationally bound at the radii given in Table 6.2 a total mass of $10^8 \ M_\odot$ is required to act at the centre. This minimum mass is in agreement with independent estimates of the total energies involved in the quasar phenomenon.

During 1966 significant absorption features were observed in the spectra of quasars, in particular those quasars with large red-shifts. Of the hundred-odd optically identified quasars, 20% were shown to have at least one absorption line. Although it is difficult to assign a red-shift on the basis of just one line, Burbidge, Burbidge, and Hoyle (1967) have made an analysis of the red-shifts associated with the absorption lines in quasar spectra and conclude that for all cases so far either $z_{absorption} \approx z_{emission}$ or $z_{absorption} \sim 1 \cdot 95$. In a few cases $z_{absorption} \sim z_{emission} \sim 1 \cdot 95$. All of these cases are unusual; they indicate (a) that the absorption and emission regions in quasars are not coincident, (b) that the red-shift $z \sim 1 \cdot 95$ has special significance either in cosmology or in the quasar model. These results are based on a comparatively small number of quasars; should subsequent observations support these conclusions then it would appear to be evidence against the cosmological interpretation of the red-shift.

6.7 Optical variations

Since the optical component of quasars is too small to be resolved, upper limits to the optical size can only be estimated from the time-scale of the variations in optical intensity. Most quasars are too faint to appear on old photographic plates; Smith and Hoffleit (1963) found records of 3C 273 on the Harvard College Observatory Sky Survey plates which date back seventy years. An analysis of these plates indicated that variations with time-scales of the order of months were superimposed on a smaller variation with a period of about ten years. 3C 48 has also been found to be variable. Several other quasars have exhibited variations since their discovery. 3C 345 varied by 40% over a few weeks; 100% increase in brightness was observed in 3C 446 between two successive nights of observations.

The emission from the latter object is strongly polarized indicating that the magnetic bremsstrahlung mechanism is involved. This mechanism is not particularly effective in the small volumes and high photon densities that are indicated for quasars unless the magnetic fields are high. Otherwise the Compton effect will drain the relativistic electron energies into gamma-rays causing the electrons to have a prohibitively short lifetime. The half-life of an electron against magnetic bremsstrahlung energy loss is given by

$$t_{\frac{1}{2}} = 4 \times 10^{14} H^{-2} E^{-1}, \tag{6.1}$$

where E is the electron energy in eV, H is in gauss and $t_{\frac{1}{2}}$ is in seconds. Higher magnetic fields cause a loss of most of the electron energy to radio and optical emission. Either way the short lifetimes indicated are difficult to reconcile with the light variations. When the electron lifetime is of the order of seconds, how can light variations be propagated through dimensions of the order of light-days? Electrons must be injected at the same time all over the quasar rather than at some central region; one mechanism by which this could take place has been suggested by Hillier (1966). Relativistic protons are injected from some central source; these undergo proton–proton interactions giving mesons which decay to electrons and gamma-rays. These collisions will occur uniformly over the quasar since proton loss by magnetic bremsstrahlung emission or Compton scattering is negligible. The optical variations will result from variations in relativistic electron densities at any moment which, in turn, reflect variations in the proton injection rate. This theory would indicate that the gamma-ray fluxes, which also arise from meson decay, should be detectable with the rapidly developing gamma-ray detection techniques. This flux should also reflect the variations in the proton injection rate.

Several ingenious ideas have been suggested to explain the short-term optical variations without restricting the quasar size. None of these can yet be considered satisfactory and the optical variations still rank as one of the many unexplained features of quasars. The observation of these variations is strong evidence for the continuing activity of the energy source in these objects. Since the periods are so short, observations are obviously required by regular surveys, placing heavy demands on the already small supply of large optical telescopes.

6.8 Radio luminosity diagrams

A number of attempts have been made to show some systematic variation between the luminosity of galaxies, radio galaxies and quasars and the other parameters of the sources. This systematic variation would enable an evolutionary pattern on a galactic scale to be recognized and the already considerable amount of data to be classified. At the moment each extra-galactic radio source must be treated on its own without the possibility of fitting the object into a general sequence of galactic evolution. Thus it is not possible to deduce whether radio galaxies are the products of a single explosion, whether they evolve through all the various forms observed or whether these forms are different evolutionary paths. The H.R. diagram gave meaning to the observational data on stellar objects by showing the general outline of the Main Sequence and hence the evolutionary pattern of stars. A similar diagram is urgently required for galactic studies; there is now sufficient data accumulated, but the meaningful parameters have yet to be decided upon. One of the attempts at such a galactic sequence diagram was that by Shklovsky (1962), who plotted radio luminosity L, versus radius R, for a number of extra-galactic radio sources. A Japanese group [Aizu *et al.* (1964)] drew up a similar diagram in which they plotted L against L/V, where $V =$ volume, for 160 optically identified extra-galactic radio sources (including ten quasars). The resulting diagram had the general form shown in Fig. 6.3. The Giant Sequence is made up of

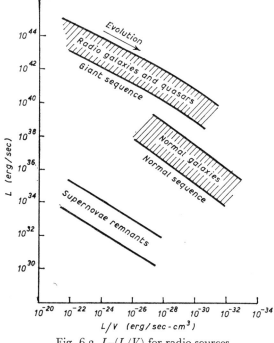

Fig. 6.3. $L-(L/V)$ for radio sources

radio galaxies and quasars. The Normal Sequence is composed of normal galaxies, such as the Galaxy, which are weak radio emitters. In each sequence the direction of evolution is from left to right. A cluster of normal galaxies, e.g., the Virgo cluster, will lie along a straight line in this diagram just as stars in a cluster evolve along one path in the H.R. diagram. An interesting feature of this representation is that if supernovae remnants are plotted on the same diagram they form a sequence parallel to the galactic sequences but displaced by 10^{10} in luminosity.

An alternative version of this diagram using more recent data has been drawn up by Heeschen (1966). The luminosity L is plotted against the surface brightness B

$$L = 4\pi D^2 I$$

and
$$B = \frac{LI}{\pi\theta^2}$$

where I = flux density at 1400 Mc/s, D = distance to source, and θ = angular radius in radians.

The flux density at a particular frequency was taken rather than its integral over a range of frequency, which had been previously used. Integration involves assumptions about the nature of the radio spectrum whereas accurate values of I at 1400 Mc/s are available for almost all extra-galactic radio sources.

The resulting diagram is plotted in Fig. 6.4; again there are two sequences. The upper sequence containing many quasars corresponds

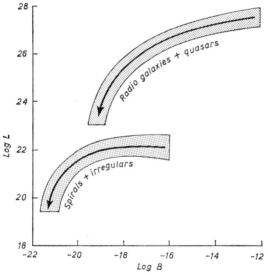

Fig. 6.4. Plot of extra-galactic radio source luminosity against surface brightness

to the Giant Sequence of the Japanese group. The evolution starts at the upper right-hand corner where the galaxy explodes as a small bright object. The volume gradually expands but the luminosity stays approximately constant until a stable volume is reached. The luminosity then falls off with time with very little change in size. The lower sequence is composed of spiral and irregular galaxies which evolve in a similar path but are reduced by 10^5 in energy. Two peculiar galaxies fail to conform to this classification; NGC 1068, a Seyfert galaxy which is a weak radio emitter, and M 82, the exploding irregular galaxy. Heeschen suggests that the position of these galaxies to the right of the sequence may be due to a secondary explosion of recent origin.

It is worth noting that quasars only conform to this picture if they lie at cosmological distances. If they were locally ejected objects, then anyway they would not belong in a system of objects of galactic size.

6.9 Spatial distribution

The cosmological interpretation of quasar red-shifts suggests that their distribution should be isotropic in most cosmologies. The verification of this property is difficult since selection effects are inevitable. Arp (1966) has made a study of peculiar optical galaxies and has found that a number of these are grouped such that it appears that two radio sources, usually quasars, have been ejected from a central object which lies between them. Some of these objects are known to lie at distances less than 100 Mpc so that an alternative explanation of red-shift must be provided. Recently the statistical significance of Arp's results has been questioned.

Most galaxies occur in clusters. The occurrence of a quasar in a cluster of galaxies would be evidence for the galactic scale of the phenomenon. Radio galaxies are often associated with the brightest member of a cluster. Most quasars are so faint optically (less than $+16$ magnitude) that other weaker galaxies in their immediate vicinity would not be detectable as such. Only 3C 273 and 3C 48 are sufficiently close that their fellow cluster members would be observable; these have not been observed. This is slight evidence against the galactic nature of quasars but confirmation must await the use of more sensitive photographic techniques to photograph the region around the weaker quasars.

A study by Strittmatter, Faulkner and Walmsley (1966) of the position of quasars in the sky shows that those with $z > 1.5$ have an anomalous distribution which may have important consequences. With the exception of one source, all quasars with $z > 1.5$ are contained in two compact regions of the sky. The quasars with $z > 1$ but < 1.5 have a distribution centred on these two regions; those with $z < 1$ are isotropically distributed. These results are based on a comparatively small number of sources but great care has been made to eliminate selection effects. If these results are borne out by subsequent observations, then there are three possible explanations: (*a*) the universe is anisotropic, (*b*) the quasars have a local origin, (*c*) the universe is inhomogeneous.

6.10 Log *N* – log *I* relation

If the cosmological interpretation of the quasar red-shifts is accepted, then quasars are of particular interest for cosmology. The largest red-shift for a radio galaxy is $z = 0.46$ (3C 295) whereas quasars can have red-shifts in excess of two. They are the farthest observable objects in the universe and hence the ones most susceptible to cosmological effects.

Burbidge (1967) has reviewed the implications of quasar studies for cosmology. In general the spread in optical and radio luminosities is greater for quasars than for radio galaxies, which complicates their study. One useful method of summarizing the data is the log N–log I diagram, where $I =$ radio flux and $N =$ the number of sources with brightness greater than I. If the sources are uniformly distributed in Euclidean space and have luminosities which are independent of distance (and, therefore, time), the slope of the curve should be -1.5.

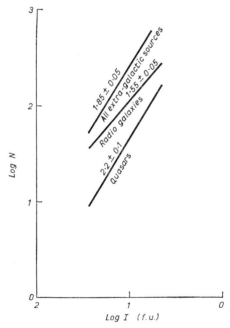

Fig. 6.5. Log N–log I diagram for extra-galactic radio sources

For very large red-shifts there are two important effects which can change this slope: (i) at greater red-shifts the brightness will appear less because of recession; this is in addition to the normal inverse square law decrease, (ii) in the past the density of sources may have been greater.

Veron (1966) has drawn up a log N–log I curve for the sources in the 3C catalogue. He finds that all extra-galactic sources give a curve with slope $-1·85 \pm 0·05$ (Fig. 6·5). This is greater than the predicted slope of $-1·5$. Since the cosmological effects (i) and (ii) would not explain this discrepancy, some other explanation must be sought.

Most of the 3C sources have been identified, so that the data can be broken up into two groups: radio galaxies and quasars. Radio galaxies have a slope of $-1·55 \pm 0·05$, while quasars have a larger slope of $-2·2 \pm 0·1$. The discrepancy is therefore associated with quasars and may be explained by (i) evolutionary effects in quasars, (ii) a local origin.

7

Extra-galactic Radio Source Theories

7.1 Energy sources

The fundamental problem facing theorists who seek to explain radio galaxies or quasars is to find the source of the very large energies released. Just as thermonuclear reactions were eventually discovered to be the source of stellar energy, so too some hitherto unknown process may be the energy source for these objects. Whatever the mechanism it appears to release efficiently significant amounts of energy, that is, a relatively large fraction of the rest mass. Most of the known processes of energy release have been proposed at one time or another for the release of energy violently on a galactic scale. So far no one mechanism can be positively identified as the basic process. Several new and novel mechanisms have been proposed, some of which, if valid, would call for a revision of previously accepted physical laws.

Burbidge and Burbidge (1965) have summarized the various energy sources postulated for radio galaxies under three headings: (a) interaction with external matter, (b) internal energy, (c) stellar-type energy source. This summary is sufficiently general to apply also to the quasar phenomenon.

The original proposal of galactic collisions comes under (a). Since galaxies occur in clusters the chances of a collision are not negligible. The double nature of many radio galaxies could be easily explained in this way. Subsequent observations of optically identified radio galaxies do not support this hypothesis since many sources appear to be explosions centred on a single nucleus. The total energy released is also difficult to explain on this model.

A more attractive proposal, from the energy standpoint, is the interaction of the galaxy with a locally concentrated region of anti-matter. Although the absence of large quantities of anti-matter is unexpected from symmetry arguments, the actual mechanism by which a large concentration could exist without annihilation is unknown. Matter-anti-matter annihilation is the most efficient energy source possible for a radio galaxy since all the rest mass is annihilated.

Accretion of inter-galactic gas by the gravitational attraction of a large galaxy could also be an important energy source. The energy accumulated is released in plasma jets such as that observed in M 87.

The internal energy of a galaxy, which is normally quiescent but is released violently under certain circumstances, is the basis of the proposals that come under (b). The formation of a proto-galaxy involves rotational, turbulent and magnetic energies, which could be released as a stable configuration is achieved. Observations show that radio galaxies are often associated with older galaxies in which these energy sources would have been released. Another proposal is that flares could occur in the large galactic nucleus in the same way as they occur in the outer layers of the sun.

The third category is probably the most important since it includes the gravitational and thermonuclear processes. The latter has already been considered for stars; it may also be important for galaxies, although it is still necessary to explain how the energy released is converted to the high-energy forms observed. The proposal that gravitational energy is released in significant quantities as galactic masses condense has been the source of much controversy. This proposal is particularly applicable to quasars. Two possibilities occur: either the energy is released in the condensation of a single massive 'star' or there is a high density of stars which interact with one another. Under certain conditions a chain reaction between adjacent stars could give rise to a greatly enhanced rate of supernovae explosions. This proposal has the advantage that it draws upon the supernovae outburst, a relatively well-understood high-energy astrophysical process, as its basic mechanism. The proposals that come under category (c) are the most important and will be briefly discussed in the following sections.

7.2 Red-shift

The deduction of large energies associated with radio galaxies and quasars is based on estimates which place these objects at very large distances from the Galaxy, sometimes at the edge of the observable universe. These distance estimates are based on the cosmological interpretation of the red-shift z associated with their spectral lines, that is, a Doppler shift due to the general expansion of the universe. The scale is established by independent estimates of the distance to nearby galaxies: it is then extrapolated to more distant objects and larger red-shifts, assuming a linear relationship, i.e.

$$d = \frac{c}{H} z,$$

where d = distance and H = Hubble's constant. Although there is some uncertainty as to the precise value of H, Hubble's Law has been accepted as the only means of establishing distances on an extra-galactic scale. For quasars the conventional interpretation of the red-shift causes so much difficulty for theoretical models that alternative interpretations have been sought which would allow the quasar to be located locally and hence decrease the total energy requirements.

Two interpretations are immediately obvious: (a) local Doppler, (b) gravitational. The local Doppler interpretation suggests that the quasars are moving with the velocities indicated by their red-shifts but that these velocities have an origin other than the general expansion of the universe. One explanation is that they have been ejected in one or more gigantic explosions of a massive object. Since blue-shifts would be expected if the seat of this explosion were more distant than the present distance of the quasars from their point of origin, it is usually assumed that the explosion occurred in the nucleus of a nearby galaxy or even within the Galaxy. The implications of this Local Theory of origin will be considered in the next section.

The gravitational interpretation of the red-shift suggests itself since quasars are concentrated objects of large mass in which gravitational effects are expected to be important. The gravitational red-shift is given by:

$$z = \frac{\Delta\lambda}{\lambda_0} \sim \frac{GM}{Rc^2},$$

where λ_0 = rest wavelength, $\Delta\lambda$ = wavelength shift, G = gravitational constant, M = mass of object, R = mean radius.

The principal objection to this interpretation comes from the width of the spectral lines; although these are broad they are only a small fraction of the red-shift. If the red-shift was due to a gravitational potential which varied radially from the star from zero to a maximum value, then the red-shift should vary from zero to some maximum value if the emission region occupied the full range of gravitational potential. The width of the spectral line therefore gives a measure of the range of gravitational field within the region of emission. In practice the line widths are only about 50 Å which means that all the emission must come from a narrow shell with a relatively uniform gravitational field. Greenstein and Schmidt (1964) have analysed this interpretation for

H

the quasars 3C 48 and 3C 273. If ΔR = thickness of emitting shell and R = radius of shell, then

$$\frac{\Delta R}{R} = \frac{\text{width of spectral line}}{\Delta \lambda}.$$

For 3C 48 and 3C 273 the values of $\Delta R/R$ are 0·016 and 0·07 respectively.

From the emissivity of the H_β line the electron density can be deduced if an electron temperature is assumed. Greenstein and Schmidt consider two cases: (1) that the quasars are stellar objects, at least 100 parsec from the sun, (2) that the quasars are massive extra-galactic objects at distances sufficiently large that they have no detectable effect on the Galaxy. In the former case (which corresponds to a neutron star) the electron density is 10^{10} times greater than the maximum possible from the observation of forbidden lines using reasonable estimates of electron temperature. In the second case, for 3C 48 a mass $M > 7 \times 10^{10} M_\odot$ with a radius $R < 0·01$ parsec is required; that such an object, which has large forces tending towards implosion, could exist with a narrow emitting shell is considered implausible.

A number of predictions have been made as to the properties of these collapsed objects. Densities from 10^{13} to 10^{19} g/cm³ are suggested; the theory of general relativity sets a limit to the value of z. Greenstein and Schmidt consider the maximum value deduced by Bondi of $z < 0·6$ as plausible. Subsequently values of z up to and greater than 2·0 have been observed; this seems to indicate that either the theory of general relativity is inadequate or that the red-shifts are Doppler in origin.

7.3 Local origin

The difficulties associated with the location of quasars at cosmological distances are largely overcome by an alternative theory known as the Local Origin theory. This was originally proposed by Terrell in 1964; it has since been developed by Terrell and others. Basically the theory suggests that quasars are the expanding remnants of a large explosion. The scale of the objects is stellar, rather than galactic, and the red-shift is a true Doppler shift. The strongest evidence in favour of the Local theory is its ability to account for a number of observational features which give rise to difficulties in any cosmological model. These points are listed below:

(i) *Energies*. The light output from a cosmological quasar is equivalent to that from 10^{12} suns or one hundred normal galaxies. From the dimensions of the radio sources the minimum age is 10^6 years. A total energy of 10^{60} to 10^{62} ergs is therefore indicated. This is too large to be explained by any known astrophysical process. On the other hand, if the quasars are local, then their optical luminosities are of the same order as that of a very bright star.

(*ii*) *Optical variations.* The size of the region of optical emission in quasars as indicated by the optical variations is difficult to reconcile with the total luminosities and the presence of forbidden lines. If the size is of the order of light-days, then the quasar dimensions are 10^{-8} times the dimensions of galaxies of equivalent power. It is improbable that the magnetic bremsstrahlung mechanism could be responsible for the optical continuum under these circumstances. In the Local theory the optical continuum is of thermal origin, the result of the usual stellar thermonuclear processes. Terrell takes the masses involved to be 10^3 to $10^4 M_\odot$, much larger than ordinary stars. The wide emission lines and the very rapid light fluctuations can then be attributed to rotation.

(*iii*) *Radio variations.* Since the radio spectrum is almost certainly due to magnetic bremsstrahlung, the recently reported variations in the microwave region raise more serious difficulties than the optical variations. Magnetic bremsstrahlung theory is sufficiently well advanced to permit a lower limit to the source dimensions to be set from the consideration of synchrotron self-absorption which occurs at very high photon densities. The dimensions indicated by this method are angular; the linear dimensions are deduced from the time-scale of the radio variations. These dimensions can then be compared and the minimum distance, compatible with both estimates, calculated. Terrell has considered two sources in detail: 3C 273B and CTA 102. His results are summarized in Table 7.1.

TABLE 7.1

Distance to sources from radio variations

SOURCE	ANGULAR DIAMETER (from synchrotron self-absorption) (seconds of arc)	LINEAR DIAMETER (from radio variations) (light-years)	COMPATIBLE DISTANCE TO SOURCE (light-years)	COSMOLOGICAL DISTANCE TO SOURCE (light-years)
3C 273B	>0·04	<20	<10^8	$1·5 \times 10^9$
CTA 102	>0·02	<0·4	<4×10^6	4×10^9

If it is assumed that the radio source size is a function of frequency, then by taking a small size for the higher frequencies, where the variations are observed, and a larger size for the lower frequencies, at which self-absorption is important, the apparent discrepancy can be resolved for 3C 273B. It is not possible to do this for CTA 102. However, the radio fluctuation observations for this source have not been confirmed so that no definite conclusions can be drawn at this stage.

In Terrell's model the radio emission arises from the continuous transfer of energy from the rapidly moving quasar to the inter-galactic gas through which it is travelling. The resulting relativistic particles radiate by magnetic bremsstrahlung and the variations are due to irregularities in the inter-galactic gas density or the inter-galactic magnetic field.

 (*iv*) *Lyman-α absorption* (see 6.2).

 (*v*) *Radio source counts* (see 6.10).

 (*vi*) *Spatial distribution* (see 6.9).

Despite these features it is not possible to accept the Local theory without reservation. The whole question of an explosion at the centre of a galaxy is speculative. There is evidence of large explosions in M 82 and NGC 1275 but these are strong radio sources. Even if gravitational implosion were to supply the energy for the ejection of the quasars, it is difficult to see how they survived the ejection process. The large hydrogen clouds observed about the Galactic centre are evidence in favour of an explosion.

The most serious objections to the Local theory are listed below:

 (*i*) *Proper motion.* From the observation of 3C 273 on old stellar plates, an upper limit can be set to its proper motion, that is, the motion relative to the frame of reference of the 'fixed' stars. This limit is less than 10^{-8} radians/year; this sets an upper limit to the distance the quasar can be from the solar system and still be moving at the velocity indicated by its red-shift. If the quasars were ejected by an explosion in the centre of the Galaxy then little proper motion would be expected since the quasars would be moving away from the terrestrial observer almost radially. If the explosion originated in another galaxy, then a minimum distance of 10 Mpc is indicated. If they originated at the centre of the Galaxy, then 3C 273 could be only 180 kpc away.

 (*ii*) *Blue-shift.* The most serious objection to the local Doppler interpretation of the red-shift is the failure to detect quasars with blue-shifts. It is more difficult to observe blue-shifts but this is unlikely to be the complete explanation. In Terrell's revised model the distance to 3C 273 would be less than that to the nearest large galaxy so that the quasars would be closely associated with the Galaxy.

 (*iii*) *Energy.* The explosion, which caused the ejection of the quasars at high velocities, must represent a gigantic release of energy. While the individual luminosity requirements of each quasar now become trivial, all the energy must be released in a single explosion of a massive galactic nucleus. A hundred objects, with masses greater than the solar mass and moving with velocities close to the velocity of light, represent a considerable investment of kinetic energy. The galactic nucleus may have a mass of $10^9 M_\odot$. If the seat of the explosion is within the Galaxy or in a nearby galaxy, then the phenomenon must be associated with 'weak' radio emitters. A similar explosion would therefore be expected in all galaxies; the radio galaxies could represent these explosions at an earlier stage.

7.4 Cosmological models

If the cosmological interpretation of the red-shift is accepted, one is faced with the problems listed in the previous section, in particular, the construction of a model which, with a lifetime of the order of a million years, can display luminosity changes over a period of weeks, and whose total luminosity exceeds that of a galaxy but whose volume is very much smaller. For a model for quasars to be satisfactory it must explain the formation, structure and eventual evolution of these massive condensed objects. The problem under each of these headings is briefly stated below; in general the models that have been proposed have concentrated on a description of the present structure.

(*i*) *Formation.* The first problem is to explain how objects with masses of this order can condense to compact volumes. The condensation of inter-galactic hydrogen, with densities less than 10^{-29} g/cm^3, into galactic-scale objects is fundamental not only to quasars but also to the formation of normal galaxies. Usually a condensing massive gas cloud will tend to fragment into smaller condensations which form the globular clusters and hence eventually stars. In most galaxies there is a galactic nucleus with mass much greater than that of stars or clusters; the role of this nucleus in galactic formation and evolution is probably fundamental but is still not understood. In the formation of a quasar this nucleus may dominate with the emission of large amounts of energy, the ejection of relativistic jets, and the formation of double radio sources. The basic problem is to explain why in some circumstances a normal galaxy may be formed, in other circumstances an explosive quasar or radio galaxy.

(*ii*) *Structure.* Having accounted for the early history of the quasar the next problem is to find a structure that is relatively stable, that is, that can survive without radical change for some 10^6 years. During this time there must be some efficient mechanism for the conversion of a considerable fraction of the total rest mass energy into observable energy forms, that is, the radio and optical continuum radiation and the spectral lines. If the former is due to magnetic bremsstrahlung, then a method for the acceleration of relativistic particles must be proposed. The radio and optical variations must be accounted for with their appropriate time-scales.

The most interesting feature of quasars comes under this heading, that is, the energy process. The possibility that some new and fundamental process might be at work has been the major reason for the attention that has been devoted to quasars.

(*iii*) *Evolution.* The third important aspect of quasars is their eventual evolution. It would be very satisfactory if quasars could be shown to evolve eventually to radio galaxies or to fit into some general pattern of galactic evolution. So far there has been little progress in this direction and it may be that quasars, radio galaxies and normal galaxies are alternate, rather than successive, modes of evolution. If gravitational forces play the important role in quasars that has been suggested, then the fate of collapsed objects is of prime importance. Whether the quasar can collapse beyond its Schwarzschild radius or approach a singularity is still an open question. Progress in this direction must come from the development of theories of gravitation since there is little hope that observations can contribute much about this final state. The possibility that large, collapsed, but invisible, masses might constitute a significant portion of the universe has been suggested.

7.5 Super-stars

About the same time that the star-like appearance of quasars was discovered and the problems associated with the large red-shifts recognized, Hoyle and Fowler (1963) published a series of papers on the possible existence of super-stars. These massive star-like objects were examined primarily to see if radio galaxies might be associated with some stage of their development; from a theoretical point of view these models were of interest even if they were not feasible in nature as they were a test of the limits of models of stellar structure. The observation of explosions in M 82 and the Seyfert galaxies seemed to indicate that the galactic nucleus might be associated with these super-stars. As the properties of quasars became apparent, it was obvious that super-stars were immediately relevant; most subsequent work on super-stars has been pursued with a view to explaining quasars.

In stellar astronomy the upper limit of the mass of stars only goes to $10^2 \, M_\odot$; the super-stars proposed by Hoyle and Fowler had masses from 10^5 to $10^8 \, M_\odot$. Although the standard techniques for the construction of stellar models can be applied to these super-stars there are some obvious differences; in particular, gravitational effects predominate. Since the theory of general relativity must be used for a rigorous solution of the gravitational effects on this scale, considerable effort has been expended in the development of the Einstein formulation to a form suitable to computation. This has not yet been achieved and the most significant results obtained so far have come from treatments intermediate between the classical and Einstein theories.

In their earlier treatments Hoyle and Fowler took as their starting point a massive object ($M > 10^4 M_\odot$) in hydrostatic equilibrium. The prehistory of the object was not considered. Hydrogen and helium burning proceed in the normal way but with the important difference that the star is entirely convective and that the radiation pressure is very much greater than the gas pressure. The mass–luminosity relationship is linear:

$$L = 2 \times 10^{38} \, (M/M_\odot) \text{ ergs/sec.}$$

For $M = 10^8 M_\odot$, $L = 2 \times 10^{46}$ ergs/sec, which is of the same order as that observed from quasars. If only half the hydrogen is burned to helium and 0·7% of the rest mass of a hydrogen atom is released, then the total energy released, E, is given by:

$$E = 1/2 \times 0·07 \, Mc^2 \text{ ergs}$$
$$\sim 6 \times 10^{51} \, (M/M_\odot) \text{ ergs}$$
$$\sim 6 \times 10^{59} \text{ ergs for } M = 10^8 M_\odot.$$

This amount of energy is just sufficient for the smaller quasars. The lifetimes deduced for the Main Sequence (hydrogen burning) stage of these objects is independent of mass and is of the order of 10^6 years. The surface temperature during hydrogen burning is more than 10,000 °K which should give strong emission in the ultraviolet continuum.

In their first paper Hoyle and Fowler followed the evolution of the super-star beyond helium burning to the point where the central temperature was 2×10^9 °K. At this stage energy loss by neutrino emission competes with the energy released by oxygen burning and a super-supernovae was predicted. In a later paper account was taken of loss of energy by electron–positron pair production as the temperature approaches 2×10^9 °K. This process is very much greater than the energy released by nuclear processes. To compensate for this energy loss, energy must come from the gravitational potential energy. Hydrostatic equilibrium breaks down and implosion occurs with velocities approaching the speed of light. If the star is spherically symmetrical then the energy released approaches the limit set by general relativity. The case in which the star has a finite rotation and hence lacks spherical symmetry is harder to solve mathematically and is probably of more practical importance. Much theoretical work has been devoted to the extension of this super-star hypothesis. The observational data do not yet enable the theory to be checked, nor is it likely that observations can be of much assistance in deciding the validity of the theory for some time to come. The mathematical complexity associated with gravitational theory is such that the importance of gravitational collapse as an energy source cannot be established. A number of authors have suggested variations on the super-star hypothesis but the basic facets of the proposal remain unaltered. Some aspects of the super-star theory are worthy of special attention.

(*i*) *Schwarzschild singularity*. In gravitational theory the accumulation of matter in a finite volume eventually leads to the escape velocity at the surface exceeding the speed of light. At this point neither particles nor radiation can leave the object. The radiation emitted at a distance R from the centre of a sphere of mass M is red-shifted to an external observer by:

$$z = \frac{1}{(1 - 2GM/Rc^2)^{\frac{1}{2}}} - 1.$$

When $R = 2GM/c^2 = R_s$, then z goes towards infinity and a singularity occurs. This radius R_s is called the Schwarzschild radius and the phenomenon is known as the Schwarzschild singularity. The time dilation factor also varies as z so that as R approaches R_s, to an external observer the motion of a collapsing surface will appear to slow down and R_s will never be reached. Signals sent by an observer on the surface at equal spaces of time in this rest frame will reach the external observer with increasing time intervals between recorded signals. Eventually no more signals will be recorded.

The Schwarzschild radius is normally not approached, as seen from Table 7.2.

TABLE 7.2

Schwarzschild radius

OBJECT	RADIUS (cm)	R_s (cm)
Proton	10^{-13}	10^{-33}
Sun	7×10^{10}	$2 \cdot 6 \times 10^5$
Super-star	10^{19}	10^{13}

Only for highly collapsed objects has this singularity any importance. Since both radiation and particles are confined within R_s, the only possible means of detecting an object which has collapsed is through the influence of the gravitational and electrostatic fields. Since these effects would not be very large, large masses could exist in the universe and remain undetected.

The red-shift suffered by most energy forms, moving out of a strong gravitational field, is a severe limitation on the effectiveness of gravitational collapse as a mechanism of energy release.

(*ii*) *Evolution of quasar*. Hoyle and Fowler have considered the possibility that a super-star might evolve to a radio galaxy. When the super-star goes into the gravitational contraction phase, the outer envelope may be ejected explosively, separating into two components moving in opposite directions with relativistic velocities. For a total mass of $10^8 M_\odot$, the ejected parts could each be of the order of $10^7 M_\odot$ with kinetic energies of 10^{20} ergs/g. This doubling feature would tally with the observed doubling of many radio sources.

Fowler (1966) has developed this concept of a super-star being composed of a core and an envelope. Gravitational implosion occurs at $R = 10^{18}$ cm, at which point the central temperature is $2 \cdot 5 \times 10^5$ °K and the central density 4×10^{-10} g/cm^3. For a spherically symmetrical object the duration of the implosion $\sim 10^3$ years; by considering rotation this period may be increased to 10^6 years. The energy released by the implosion of the core is transferred to the envelope which explodes. Because of rotation the ejected matter will stream out into inter-galactic space from the poles.

The chief difficulty with this model is to explain the transfer of the implosion energy of the core to the envelope. Fowler suggests gravitational radiation as a possible mechanism. This process is estimated as only 40% efficient. For a system in which the mass of the core $M_c \sim 0 \cdot 3 M$, where M is the total mass of the system, the energy given out by the core is about $0 \cdot 02 Mc^2$. The total energy taken up by the envelope is then given by:

$$E = 8 \times 10^{-3} Mc^2$$
$$\simeq 10^{60-62} \text{ ergs for } M = 10^{8-10} M_\odot.$$

Under similar conditions the thermonuclear energy release is about $10^{-3} Mc^2$.

(*iii*) *Relaxation oscillations*. Fowler has proposed that the dynamic instability predicted by the theory of general relativity could lead to relaxation oscillations in non-rotating stars with $M \sim 10^6 M_\odot$. These oscillations are the result of the balance between hydrogen burning and the gravitational forces tending to implode the quasar. This pulsating may represent the quasar stage, in which a central object pumps energy into an extended shell from which the optical and radio emissions are observed. Eventually the nuclear fuel will be exhausted, and collapse will occur with evolution perhaps to a radio galaxy. This model was computed using the post-Newtonian approximation to gravitational theory.

7.6 Star clusters

A number of theories have been proposed in which the large condensed mass exists as a number of discrete, but gravitationally bound, masses, rather than as a single massive object. These theories have the obvious advantage that they avoid the necessity of treating the properties of objects on a scale hitherto unexplored; the discrete masses are of the order of stellar masses and stellar models are directly applicable. The interaction of the stellar objects with one another is complex but a number of mechanisms have been proposed whereby the large luminosities observed might be generated. These include stellar collisions, chains of supernovae explosions, and thermonuclear reactions in the stars. The most striking evidence in favour of this approach is that it suggests that quasars have basically the same form as normal galaxies. It is much easier to fit such a theory into a general theory of galactic formation and evolution. A large mass of gas of the order of 10^{11} to $10^{12} M_\odot$ begins to condense out of the inter-galactic gas. As the condensation proceeds under its own gravitational forces, local instabilities occur. Eventually these instabilities are sufficient to cause the total mass to disrupt into two or more masses, which are bound to one another and which continue to contract. These individual masses become unstable and subdivide again. In this way the large diffuse cloud of gas eventually breaks down into masses of stellar magnitude. In the process some condensation may be sufficiently energetic to allow part of the mass to be ejected. The magnetic fields between the stars become highly irregular and acceleration of electrons may take place.

The details of the interactions of the separate masses with one another is complex and is unlikely to be solved in the near future. Nevertheless because of the similarity of this approach to the general problem of galaxy formation this is a fruitful field of exploration.

Hoyle and Fowler (1967) have proposed a variation of this approach in which they revive the suggestion that the observed red-shift could be gravitational in origin. Pointing out that the Greenstein and Schmidt treatment, which caused this gravitational explanation to be put aside, only applies to models where the spectral line emitting regions are on the outside of the quasar, they propose a model which claims that the emission lines come from a gas cloud at the centre of the object. In order that these lines should be visible, they assume that the bulk of the mass of the quasar is in the form of neutron stars. For $M \sim 10^{12}$ to 10^{13} $M_{\odot}$ and $R \sim 10^{18}$ cm most of the light can escape. The emitting region at the centre lies at the bottom of a gravitational potential well so that all of the lines suffer the same gravitational shift in leaving the quasar. The volume of the region of emission can therefore be considerably greater than that considered in the shell treatment of Greenstein and Schmidt. Part of the red-shift could also come from the cosmological expansion since the quasar would still be at very great distances; hence red-shifts greater than 2·0 could be explained. The absorption lines could also have a different red-shift and the missing mass in the universe could be attributed to these objects. Because of the large mass involved one would not expect to see other galaxies associated with them.

7.7 Anti-matter

Most of the physical universe with which we are acquainted is composed of matter. From symmetry arguments there seems no reason why the universe should not have been composed of anti-matter. This universe would have almost identical properties with that with which we are familiar. The possibility that half of the universe is composed of anti-matter was proposed many years ago; there are no observable facts to allow this suggestion to be rejected. It is unlikely that there is significant anti-matter in the solar system or even in the Galaxy. There is no reason why some of the more distant galaxies might not be comprised entirely of anti-matter.

If equal quantities of matter and anti-matter constitute the intergalactic gas, then the condensation of a gas cloud would proceed as before since the same gravitational forces apply to each. When particles and anti-particles come close enough together they annihilate with the emission of two gamma-rays. Since the annihilation of matter–anti-matter is the most efficient means of converting particles to electromagnetic radiation it is worth considering the relevance of this phenomenon to quasars. In the initial stages the density of nucleus will be so small that the rate of annihilation is small. Eventually the density is such that significant energy is converted to radiation. This radiation, in the form of GeV gamma-rays, must be degraded to lower energy forms. If the object could be stabilized in that the rate of energy output remains approximately constant over 10^{6} years, then a mass of only 10^{6} $M_{\odot}$ would be required for the smaller quasars.

8

X-ray Astronomy

8.1 Introduction

Until comparatively recently it was customary to treat x-ray and gamma-ray astronomy together; several reviews have summarized the astrophysical data over the complete range from keV photons to 10^{20} eV gamma-rays. Since this represents seventeen decades of the electromagnetic spectrum and encompasses many different sources and detection techniques, some subdivision is called for. This is particularly appropriate since x-ray astronomy has produced many positive and unexpected results and has become an astronomical discipline in its own right, almost ranking in importance with the older disciplines of optical and radio astronomy. Despite the increased activity in gamma-ray research, the results are still disappointing, with no hint of the surprising discoveries of its lower energy counterpart.

For this reason x-ray astronomy will be considered first, where the x-ray region has been defined for photon energies from 1 to 500 keV. All photon energies greater than 500 keV will be considered in the next chapter which will be more appropriately titled 'gamma-ray astrophysics'. The terms 'x-ray' and 'gamma-ray' will not be used here to indicate anything other than the energy of the photons, since the mechanisms involved, nuclear or atomic, are still speculative.

8.2 Discovery of the first discrete source

The first hint that x-ray studies above the atmosphere might provide surprising results came almost by accident. Although the sun had been known for some years to be a source of soft x-rays it was not thought likely that other extra-terrestrial sources would be easily detected. In June 1962 two groups from Cambridge, Massachusetts, performed a collaborative experiment at the White Sands Missile Range, New Mexico, in which they attempted to study the fluorescent x-rays from the moon's surface due to the impact of solar x-rays [Giacconi *et al.* (1962)]. The predicted flux was about 1 photon/cm²/sec which was well within the sensitivity of the detectors used.

Because of the absorbing effect of the earth's atmosphere the detector was carried aloft by rocket and exposed for a brief period above the atmosphere. The detector consisted of three large Geiger counters with mica windows of different thicknesses; the range of wavelengths, to which the counters were sensitive, was between 2 and 8 Å. The window area of each counter was 20 cm². Each counter was surrounded by a scintillation counter which was used as an anti-coincidence shield. This rejected most of the charged cosmic rays which constitute the background above the atmosphere.

The apparatus was carried in an Aerobee rocket, which remained at an altitude greater than 80 km for 350 seconds. An optical aspect system indicated the longitudinal axis of the rocket; the rocket was spun about this axis at a rate of 2 rev/sec. The normal to the counter windows made an angle of 55° with the longitudinal axis so that a wide area of sky was surveyed. The output of the counters showed a strong spatial anisotropy which was not coincident with the position of the moon. This surprising result indicated the existence of a strong extra-terrestrial source of x-rays. Two alternative explanations were considered:

(i) The anisotropy might be due to some local effect such as the earth's magnetic field or a radiation belt. Although charged particles entering from the sides of the counter were eliminated by the anti-coincidence shield, they could still enter the counter through the mica windows and be absorbed, so that the shield counter was not triggered. Since the direction of the earth's magnetic field at this point was almost coincident with the direction of the anisotropy, the magnitude of the effect that this might have on the counting rate was calculated carefully. From the sharpness of the observed peak, the pitch angles of the particles would have to be so small that the source of the charged particles would have to lie close to the earth. A source in that position is unlikely.

(ii) If the detectors were ultraviolet sensitive, the anomaly could be explained. The mica windows were coated black to prevent the counters responding to ultraviolet or visible light. The absence of a peak when the moon was in the field of view demonstrated the effectiveness of this coating.

It was therefore concluded that a discrete source of x-rays had been detected; since its location did not coincide with any object in the solar system its origin was either Galactic or extra-galactic. The intensity of the flux was 5·0 photons/cm² sec; this flux was superimposed on a background, which was probably x-rays, of intensity 1·7 photons/cm² sec sterad. The location of the source could not be fixed to more than 10°, but was quite close to the Galactic equator. Since it lay in the constellation of Scorpius, it was called SCO XR-1.

8.3 Detection techniques

(A) DETECTORS

The detectors used initially were refined versions of those which, for some years, had been used in the laboratory to detect x-rays. Because of the particular needs of the astronomical situation, that is, comparatively low photon fluxes in the presence of large fluxes of charged particles, some special detectors were developed [Giacconi (1966)]. The general requirements of an x-ray detector are that it should have a sensitive area from 10 to 1000 cm², be highly efficient and have a well-defined energy response.

(i) *Gas counters.* The first detectors used were gas-filled Geiger counters with densities sufficient to ensure that there was a good chance of photoemission from a gas molecule by the incident x-ray. The major parameters of this type of detector are the thickness, t, and absorption coefficient, $\mu(\lambda)$, of the window and the thickness, t_g, and the absorption coefficient of the gas, $\mu_g(\lambda)$. For registration, the x-ray photon must traverse the window but be absorbed by the gas; the efficiency of detection of an x-ray photon of wavelength λ is given by:

$$\varepsilon(\lambda) = e^{-\mu(\lambda)t}(1 - e^{-\mu_g(\lambda)t_g}).$$

The detector is designed so that $\varepsilon(\lambda) \sim 0$ except over a narrow band of wavelengths (Fig. 8.1). Since the window must be a finite thickness to support the pressure difference between the gas inside the counter and the near-vacuum outside, detectors are normally limited to photon energies greater than 1 keV; using special windows this lower limit can be extended to 0·4 keV. The upper limit is set by transparency of the gas in the counter; using xenon, photons up to 50 keV can be detected.

The counters may also be operated in the proportional region by using lower voltages. In this region the output pulse is proportional to the number of ions produced, which, in turn, is proportional to the energy of the x-ray. A pulse-height analysis gives the photon energy spectrum, if correction is made for the binding energy of the photoelectron. For small photon energies the fluctuations in the number of ions produced is the effective limit on the use of this technique.

(ii) *Scintillation detectors.* Thin sheets of scintillation material are more

effective than gas counters, particularly at higher energies, because of
their greater density. The photoelectron which is emitted as a result of
the collision of the x-ray photon with the scintillator material will be
absorbed. The number of light photons emitted before the electron is
captured is proportional to the energy of the electron and hence of the

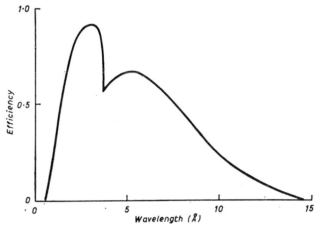

Fig. 8.1. Efficiency of gas counter with mylar window (0·88 mg/cm²)

x-ray photon. These light photons are detected by a photomultiplier;
the threshold for detection is that there should be enough light photons
incident on the photocathode of the photomultiplier to cause the
emission of at least one photoelectron. The threshold x-ray photon
energy is about 1 keV; below this energy the efficiency of detection is
low, due to the noise inherent in a photomultiplier. In practice scintillation detectors are used to detect photons with energies greater than
10 keV because of the poor energy information below this energy. In
this higher energy region the scintillation counter is superior to the
proportional counter in both energy resolution and detection efficiency.

(iii) Photoelectric devices. The use of a photomultiplier, without a
scintillation crystal, to detect x-rays was made possible by a discovery
by a group of Russian scientists in 1960. The efficiency of photoelectron
emission from a solid surface of x-ray photons in the 1 to 10 Å region
is normally less than 1%. Lukirsky and his colleagues discovered that
some alkali halides (for example, CsI, KCl) had a very high efficiency
for x-ray photoelectron emission. For photons of energy 0·8 keV the
efficiency is about 25%.

This principle has been utilized in detectors with sensitive areas of
40 cm²; no windows are required for astronomical work since the
vacuum in which the detector is exposed is ideal for operation. The

photoelectrons emitted from the sensitive alkali halide area are electrostatically focussed to the first dynode as in the conventional photomultiplier. The absence of a window and the high efficiency at long wavelengths make these detectors particularly suitable for low-energy work.

None of the detectors listed above are uniquely sensitive to x-rays. If the output data are to be of use, precautions must be taken to ensure that all sources of background are either eliminated or reduced to a reasonable level. For point source identifications the presence of a spurious component amongst the detected events is not serious as long as it is less than the background x-ray contribution. For absolute measurements on the diffuse background or on the energy spectra of discrete sources this spurious component must be rejected. The most serious source of background is charged particles, the primary cosmic radiation and its albedo. Gas counters are usually sunk in a well in a large scintillation detector which is operated in anticoincidence. Charged particles, which enter from the sides, are detected by the scintillation counter and rejected; they may still enter through the window and, if their energy is low enough, be absorbed by the x-ray detector, and hence not trigger the anticoincidence shield. This contribution is unavoidable but can be minimized by the choice of x-ray detector.

Ultraviolet photons, which would normally be transmitted by the window, may be rejected by the choice of a window which is opaque to the u. v. Scintillation counters must be similarly guarded. Since the responses of scintillation materials are different for photon and charged particles, two scintillation counters composed of different scintillation materials can be used to identify and reject charged particles.

In photoelectric devices a thin window is employed whose function is to shut out visible and ultraviolet light. This window can be sufficiently thin that the sensitivity of the detector to low-energy x-ray photons is not impaired.

(C) COLLIMATORS

A detector with a window has some degree of angular resolution if only of the order of π steradians. The angular resolution can be improved by the use of absorbers in front of the window to restrict the field of view. Collimators, in the form of a honeycomb of long metal cylinders, have been used to give a field of view of the order of a degree. Since x-ray photons are easily absorbed, the total weight of the collimator can be quite small. The response curve for a point source is usually triangular in shape, with maximum along the axis of the cylinder. (Fig. 8.2).

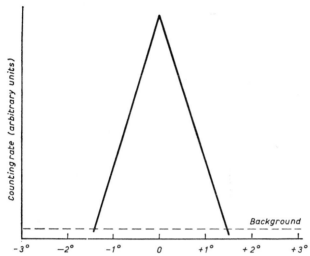

Fig. 8.2. Response of honeycomb collimator to transit of point source

This type of honeycomb collimator has the very great disadvantage that the area of sky surveyed is small, a serious consideration in rocket flights which last only a few minutes. Their principal advantage is that the analysis of data is straightforward, that is, there are no side lobes.

A more sophisticated collimator has been developed by the A.S.E.–M.I.T. groups. This consists of two grids of parallel wires separated by a small distance and mounted directly in front of the detector window (Fig. 8.3). The gaps between the wires are about the same order as the diameter of the wires, which are thick enough to absorb incident x-rays. Maximum transmission for a parallel beam of X-rays will occur only if the 'shadow' cast by the first grid falls on the wires of the second (inner) grid. This will occur at a number of angles. At orientation midway between these angles of maximum transmission the two sets of grids appear as an effective shield of the detector window and the response falls off. With this arrangement very high angular resolution can be achieved while a large area of sky is surveyed. The analysis is difficult, particularly if a complex of sources is under investigation. Even for a single source the results are ambiguous since there are many 'main' lobes.

If D = distance between grids and d = separation and diameter of grids, then a point source which transits through the field of view of the instrument will produce a modulated counting rate, with an angular distance of d/D between maxima (Fig. 8.3). For a source of finite dimensions this modulation will be modified and the source dimensions can be estimated. Angular resolution of the order of a minute of arc can be achieved by this method.

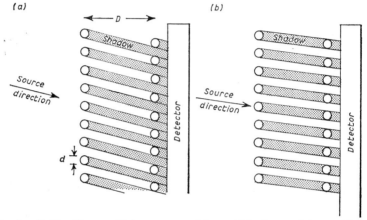

Fig. 8.3. Grid collimator. Point source positions which give a minimum (a) and maximum (b) response from the detector

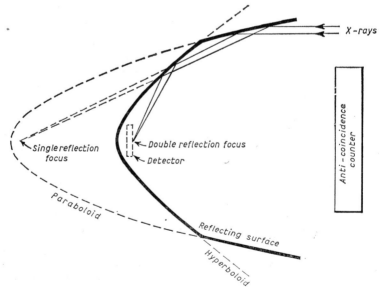

Fig. 8.4. Reflecting x-ray telescope

Both optical and radio astronomy would have been severely restricted if use could not have been made of large reflecting surfaces to increase the collection area of the instruments and hence lower the threshold sensitivity. Unfortunately x-rays cannot be easily reflected. A reflecting instrument, which forms an x-ray image, has been constructed; use is made of the fact that x-rays at grazing incidence are efficiently scattered. This scattering, or reflection, takes place at the inner surface of a cylinder in which the x-ray beam is parallel to the axis of the cylinder (Fig. 8.4). Two reflections are required to form an image. The inner surface of the cylinder is shaped so that the initial reflection is from a parabolic surface. The second reflection comes from a hyperbolic surface so that an image is formed in the plane perpendicular to the cylinder axis. These instruments are potentially the most useful development in x-ray detection techniques.

The angular resolution of a detector carried in a moving rocket, balloon or satellite is only as good as the accuracy with which the direction of the detector axis is known. This is usually determined by an optical sensor which indicates the position of the detector relative to a bright star or the sun or moon.

(D) DETECTOR TRANSPORT

Despite the high yield of interesting results from rocket and balloon x-ray experiments, satellites have not yet been used in x-ray astronomy beyond the solar system. This is largely because of the inflexibility of space programmes and the long preparation time of satellite experiments. Since the x-ray results are among the most exciting results to come from space technology, this is particularly unfortunate.

Balloons are a well-tried and relatively inexpensive means of instrument transport. Using the largest balloons, instrument packages weighing 500 lb can be carried to altitudes of about 40 km. It is not possible to get completely free of the atmosphere; a typical depth is 4 g/cm², with the depths of 2 g/cm² being reached in some experiments. (The atmospheric depth at sea level is 1033 g/cm².) Balloon exposures last from 8 to 24 hours, the upper limit being set by the balloon descending at sunset because of the gas cooling. The balloon technique is not suitable for instruments with high angular resolution because of difficulty in pointing the gondola. Because there is still a significant amount of atmosphere above the detector for low-energy x-rays, the technique is best suited for the higher energy regions using instruments with large fields of view.

At lower energies rocket-borne instruments have made the major contribution. Aerobee sounding rockets can reach altitudes of 200 km; usually observations are taken above 80 km. The duration of the flight above this altitude is about five minutes. It is remarkable that the great bulk of the data now available on x-ray discrete sources and background has been accumulated from a few hours of exposure time.

Three methods of controlling the rocket motion during flight are generally employed:

(*i*) *Free spin.* This is the simplest method to employ since the rocket is spun during launch and retains the spin during the flight. This causes the detector rapidly to sweep out a wide area of sky; the rate of spin is about 2 revs/second. In addition to spinning there is precession, so that the superposition of successive scans, that is necessary to identify the discrete sources, is difficult. For preliminary surveys over wide areas this method is particularly simple and useful.

(*ii*) *Slow spin.* The rocket can be despun in flight, using gas jets. The rate of spin can be reduced to 4 revs/minute so that wide areas can be covered for long periods without having to superimpose successive scans.

(*iii*) *Controlled pointing.* For some experiments it is necessary to survey just one region of the sky. In these circumstances it is wasteful to sweep wide areas. The altitude of the rocket can be controlled by the use of gas jets which are fired on command from pre-set gyroscopes. Pointing accuracy of 1° can be achieved so that a particular source can be studied for the entire flight if the field of view of the detector is greater than this. The most remarkable use of this technique was the lunar occultation experiment on the Crab Nebula, in which the source TAU XR-1 was observed at the precise moment that it was occulted by the moon.

8.4 Early observations

Most of the x-ray results at low energies have come from observations n rockets by three groups of experimenters in the United States; the combined American Science and Engineering, Inc., and M.I.T. group in Massachusetts, the Lockheed group and the Naval Research Laboratory group. The strong competition among these groups has led to a rapid development of the field but also to a difficulty in the interpretation of the results which are sometimes conflicting and often unconfirmed. In particular the nomenclature of the sources has not yet been standardized; here the sources are named from the constellation in which they occur, followed by the letters XR and a number to indicate the order in which the sources were discovered in that constellation, for example CYG XR-2.

The discovery of the first and strongest discrete x-ray source SCO XR-1 has already been described. In this flight there was also some evidence for a source in Cygnus. In the twelve months that followed the A.S.E.–M.I.T. group confirmed the existence of the Cygnus source, CYG XR-1 using somewhat more sophisticated detectors. They also detected a weak source in Taurus, TAU XR-1. This source was particularly interesting as the Crab Nebula lay in that region of the sky. Since the angular resolution was of the order of 10°, any one of a million visible sources could have been the x-ray source.

In 1963 the N.R.L. group independently established the existence of SCO XR-1. This group had been working in the field of x-ray astronomy prior to the A.S.E.–M.I.T. flights. They used a detector with a hexagonal collimator which limited the field of view to 10°. The detector was a proportional counter of 86 cm² area which was sensitive to 1 to 8 Å x-rays. During the four-minute flight the entire sky above the horizon was scanned at least once. The Scorpius source was scanned eight times so that its position was located to within 1°. Since the Galactic centre was below the horizon during the flight it was obviously not associated with SCO XR-1. In the same flight the position of the Taurus source was confirmed to lie near the Crab Nebula; its intensity was one-eighth that of SCO XR-1.

By the end of 1963 the first phase of the development of x-ray astronomy could be considered complete. The existence of three extra-terrestrial discrete sources was established: SCO XR-1, CYG XR-1, TAU XR-1. The uncertainty in the source positions was too great to permit identification with any known astronomical object. This situation was analogous to that which occurred in the early days of radio astronomy prior to the optical identification of Cygnus A.

The second phase of astronomical x-ray research consisted in: (a) the improvement of detector sensitivity to find new sources, too weak to be detected in the preliminary surveys, (b) the refinement of detector techniques to determine source positions sufficiently accurately for optical identification, (c) the determination of the energy spectrum of the stronger sources using energy-sensitive detectors. The results of these developments will be considered in the following sections.

8.5 Position of sources

Since 1963 the number of known x-ray sources has increased rapidly. The precise number is a matter of definition, since many sources have been located in just one experiment. In some cases sources appear in one survey of a region and not in another of the same region. This could be due to variability in the x-ray luminosities, instrumental defects or statistical fluctuations.

The first ten sources discovered were grouped along the Galactic plane. The only exception was SCO XR-1 which is only 20° from the Galactic equator and near enough to make a Galactic origin not unlikely. At an early stage it was believed that all the sources lay in the Galaxy and that they were probably remnants of supernovae. This picture had to be revised after a flight by the N.R.L. group in April 1965 [Bryam *et al.* (1966)]. Two large Geiger counters of total area 450 cm² were used. A honeycomb collimator defined a field of view of 8°; an angular resolution of 1·5° was claimed. The overall sensitivity of the instrument was a factor of four greater than previous experiments. In addition the rocket was made to spin slowly (4 revs/minute). The increased sensitivity was sufficient to produce some startling results.

The most surprising result was the variability of sources in the Cygnus region. As a result of the decrease in the strength of CYG XR-1, a source CYG XR-3 was seen close to Cygnus A. This was thought to be the first evidence for an extra-galactic x-ray source. A later flight by the A.S.E.–M.I.T. group showed that this source was not coincident with Cygnus A.

Two other sources were located which lay well away from the Galactic plane and which could be extra-galactic. One of these, VIR XR-1, is apparently associated with Virgo A (the Jet Nebula) to the accuracy of the experiment (about 1°). The other, LEO XR-1, does not appear to have any well-known radio source within its circle of error.

Further evidence for extra-galactic x-ray sources has come from a group from the Goddard Space Flight Centre who have used balloon-borne instruments to detect higher energy x-rays. At the altitude of the flights the atmospheric depth was 2·7 g/cm². The basic detector was a caesium-iodide scintillation crystal which was sensitive to 20 to 100 keV x-rays. This was surrounded by an anti-coincidence scintillator. The window to the detector was a proportional counter, which was more sensitive to charged particles than high-energy x-rays, and was therefore operated in anti-coincidence. Above this counter was a copper disc; an annular gap in this disc defined the field of view. A point source passing through the field of view of the detector gives a structured response; the angular resolution is about three degrees. In a preliminary flight the source CYG XR-1 was observed. A later flight showed evidence for a broad source, or group of sources, in the region of the sky in which the Coma cluster of galaxies is centred.

At least thirty x-ray sources have now been reported (Fig. 8.5).

Twenty-five of these sources lie within $\pm 15°$ of the Galactic plane. Three of them could be identified with supernova remnants—the Crab

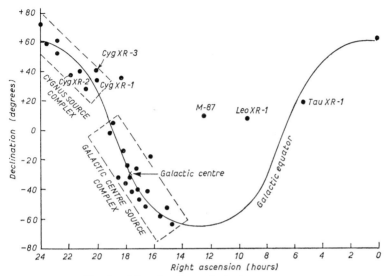

Fig. 8.5. Map showing principal x-ray sources

Nebula, Cassiopeia A and SN 1572. One survey reported a signal from SN 1604. Comparison of the position of the x-ray sources with the positions of supernovae reported by early oriental astronomers shows that at least four other sources could be associated with supernovae remnants.

The number of sources is now sufficient to enable some meaningful conclusions to be drawn about their spatial distribution. Friedman, Byram and Chubb (1967) have noted that the sources fall into two main groups. Nine sources (excluding Cass A) lie in the Cygnus-Cassiopeia region within $\pm 7°$ of the Galactic equator. Fifteen sources (excluding SCO XR-1) lie within $\pm 3·5°$ of the Galactic equator in the general direction of the Galactic centre (Sagittarius).

These distributions have been compared with (i) supernovae, (ii) novae, (iii) the spiral arms. No association is found for (i) or (ii). The grouping in the Cygnus-Cassiopeia region, and the Sagittarius region can be directly associated with the position of the spiral arms, determined by H II radio emission and the distribution of O and B stars. This association confirms the Galactic and stellar nature of the majority of x-ray sources.

8.6 Optical identification of x-ray sources

The first optical identification of an x-ray source used the lunar occultation technique which is familiar in radio astronomy. This technique is much more difficult for x-ray sources since the observations must be carried out above the atmosphere. Although the detectors available at the time were only good to a few degrees, the N.R.L. group [Bowyer *et al.* (1964)] managed to fix the size and position of TAU XR-1 to one minute of arc. They gambled on the x-ray source being coincident with the Crab Nebula which lay within the error circle of the x-ray location. The detectors used were conventional in design. The rocket launching was timed so that the detector would be above the atmosphere for the critical moments that the moon passed in front of the centre of the Crab Nebula (Fig. 8.6). The altitude of the rocket was controlled so that for 12 critical minutes this region was

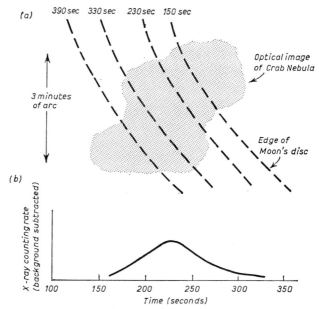

Fig. 8.6. Lunar occultation of Tau XR–1. (*a*) Position of moon's disc relative to Crab Nebula in seconds of time after launch; (*b*) the deduced x-ray source distribution, in seconds of time after launch

monitored. The experiment was a complete success and produced two important results: (i) TAU XR-1 was identified with the supernovae remnant, (ii) the size of the source was finite (about two minutes of arc in diameter). This latter result had the consequence that the neutron

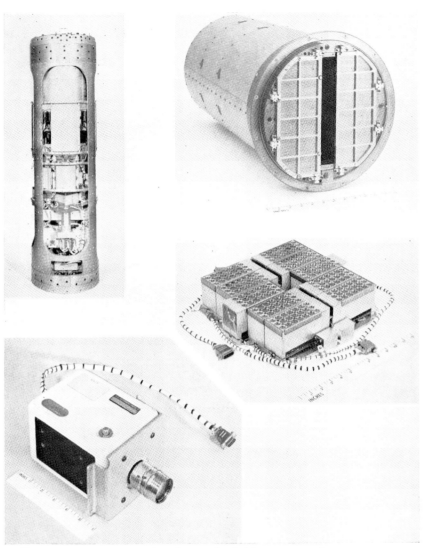

5. The x-ray detector used to locate the brightest extra-solar x-ray source, SCO XR–1. *Top left:* Complete payload carried by the Aerobee rocket. *Top right:* The two sets of modulation grids which gave the unprecedented accuracy in the location of the source. *Bottom left:* The aspect camera which gave the position of the rocket during flight. *Centre right:* The proportional counters which were mounted behind the modulation grids.

(Courtesy of A.S.E.)

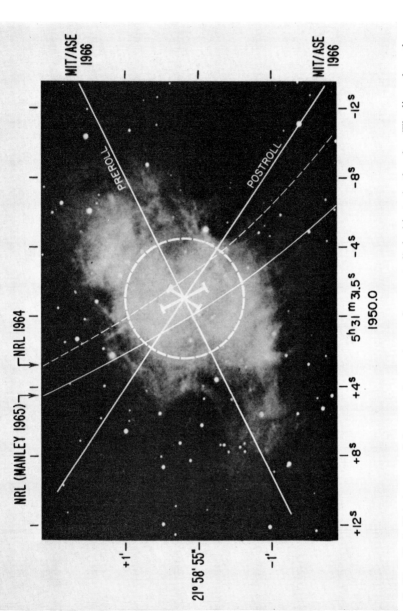

6. The Crab Nebula, one of the most interesting and most studied objects in the sky. The lines show the position of the centre of the x-ray source deduced from various experiments. The 100″ diameter circle represents the approximate size of the x-ray source. At radio wavelengths the source is approximately 4′ in diameter and centred on the optical image.

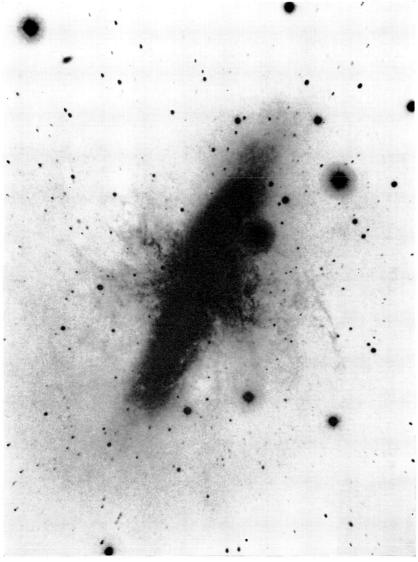

7. M82, the exploding galaxy. The rapidly moving filaments are evidence for a recent explosion and are strongly polarized. This object could be an important source of cosmic rays.

(*Mount Wilson and Palomar Observatories*)

8. The first Princeton background radiometer in its observing position on top of the Geology building at Princeton. This instrument was used to make the difficult measurement of the background 3 °K microwave field at 3·2 cm.

star hypothesis, which proposed that the source should be point-like, could be immediately rejected.

Lunar occultations are somewhat fortuitous events and cannot, therefore, be relied upon as the sole means of accurately locating x-ray sources. Fortunately the angular resolution of x-ray techniques has been sufficiently developed that positions can be fixed to a few minutes of arc without the moon. The first optical identification achieved in this way was that of the strongest source, SCO XR-1. The original discovery of SCO XR-1 only located the source in a circle of error of diameter 20°. The later N.R.L. experiment reduced this uncertainty to 1°. This uncertainty was too large to permit an identification although it was sufficient to show that this strong x-ray source was not associated with (a) a strong radio source (b) visible nebulosity. The size of the source was shown to be less than seven minutes of arc.

In March 1966 a stabilized rocket flight by the A.S.E. group [Oda et al. (1967)] succeeded in locating the source with the accuracy required for identification. Two proportional counters, of sensitive area 120 cm² each, were used to detect 1·5 to 30 keV x-ray photons. Collimation was provided by the modulation grid arrangement. Each counter had a slightly different grid so that by simultaneous observation of the same region of the sky some of the ambiguity inherent in this technique could be removed. A diffuse light source was mounted in front of the grids; this was viewed through the grids by an aspect camera which continuously photographed the night-sky. The precise part of the sky visible to the x-ray detectors at any moment was therefore recorded. The aspect of the rocket was controlled during the flight so that SCO XR-1 was observed for 55 seconds. To check the accuracy of this method of position recording, the attitude of the rocket was changed so that TAU XR-1, whose position was well known, was also observed.

This experiment had two important results. (i) The size of the source was shown to be less than 20 seconds of arc. Since the source was believed to be relatively close, the small size indicated that the object was most likely stellar in scale, rather than an extended object, like a hot gas cloud. (ii) The position of the source was located to an accuracy of ±4 seconds of time in right ascension and ±30 seconds of arc in declination. Because of the ambiguity of the collimator grid technique this error rectangle could be centred on either of two positions separated by five minutes of arc. This positional accuracy was an increase of two orders of magnitude on previous experiments.

Using these positions, optical astronomers at the Tokyo Observatory photographed the regions of interest with their 74-inch reflector. It had been predicted that the optical object would be star-like, have an apparent magnitude of at least $+13$ and have an ultraviolet excess. The brightest object in either of the two error rectangles on the Japanese plates was an object which satisfied all these criteria. The same object was subsequently observed with the 200-inch Mount Palomar telescope. These observations showed that the stellar object had many of the characteristics of an old nova in its quiescent phase. The optical magnitude varied by one magnitude over three nights of observations. From the study of the Harvard Sky Survey plates over the past seventy years it appears that there has been no outburst from the nova, if it is such, over that period.

Independently of these observations and using only the position available prior to the A.S.E. flight, Johnson and Stephenson (1967) made a survey of the stellar field in the error circle of SCO XR-1. Although this was a large area containing many stars, they came to the same conclusion as to the most likely optical object as the A.S.E. group using their precise position.

Using this information on the optical characteristics of SCO XR-1 it was possible to identify CYG XR-2 with a similar star-like object. The position of the x-ray source was fixed to a few minutes of arc by a flight which did a detailed survey of the Cygnus region. Although there were many bright optical objects in the error rectangle of CYG XR-2 only one somewhat dim star had the optical characteristics of SCO XR-1. This is not the brightest object in this rectangle, but the optical properties are sufficiently striking for this identification to be treated with confidence. There is evidence that this source is a member of a binary system (two closely associated stars).

Now that x-ray detection techniques with high angular resolution have been developed and the optical characteristics of some of the objects recognized, the number of identifications can be expected to rise rapidly. In some cases there may be considerable difficulty; CYG XR-1, although positioned to the same accuracy as CYG XR-2, lies in a region of dust obscuration. The optical object may thus be totally obscured. The definite identification of the strong radio sources, which lie in the error circles of several x-ray sources, e.g. Virgo A, Cass A, is of particular interest.

8.7 Source spectra

At this stage the most important evidence for the mechanism of the x-ray emission comes from the shape of the energy continuum spectrum. Unfortunately only spectra for the strongest x-ray sources are yet available. Much effort has been devoted to the development of detectors with high angular and energy resolution. Most of the early rocket experiments used some crude method of spectral analysis such as two identical counters with windows of different thickness. The most useful information has come from balloon experiments which determine the spectrum above 20 keV from sources whose positions have been determined in rocket experiments. Generally three types of spectrum are expected; the form of these spectra are listed in Table 8.1, together with the radiation mechanisms that would give this spectral distribution.

TABLE 8.1

Spectral types

TYPE	FORM	MECHANISMS
Power law	$I(\nu) \propto \nu^{-\alpha}$	Magnetic bremsstrahlung Compton
Exponential	$I(\nu) \propto e^{-h\nu/kT}$	Free-free emission from optically thin plasma
Thermal	$I(\nu) \propto \dfrac{\nu^3}{e^{h\nu/kT} - 1}$	Optically thick at temperature T

In an analysis of the early results from a number of surveys of SCO XR-1, Giacconi has concluded that each of the spectral forms could fit the observations in the 1–40 keV region if the parameters have the values in Table 8.2.

TABLE 8.2

X-ray source parameters (SCO XR-1)

FORM	PARAMETER
Power law	$\alpha = 1\cdot1 \pm 0\cdot3$
Exponential	$T = (3\cdot8 \pm 1\cdot8) \times 10^7$ °K
Thermal	$T = (9\cdot1 \pm 0\cdot9) \times 10^6$ °K

The strongest source at the lowest energies, SCO XR-1 has a soft spectrum (Fig. 8.7) so that at 50 keV the flux is less than that from TAU XR-1. The spectrum of the latter source was determined in a balloon experiment by Clark (1965) in India in July 1964. Using a scintillation counter as detector, pulses were sorted into four channels between 15 and 60 keV. The spectrum was consistent with a power law with exponent 2. A rocket experiment by the N.R.L. group at lower energies (1·2 to 4·0 keV) had indicated a power law with exponent 1·4.

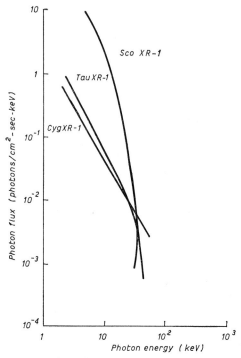

8.7. Spectra of the first three x-ray sources discovered

Other balloon experiments have shown that CYG XR-1 has a somewhat harder spectrum than TAU XR-1. An exponent of 1·7 is the best fit to the data. This is the hardest x-ray spectrum yet discovered and a likely source for gamma-rays.

The observed spectrum of SCO XR-1 can be best represented by an equation of the form

$$n(E) \propto \frac{1}{E}\, e^{-E/kT}$$

which is characteristic of thermal bremsstrahlung with $kT = 4\cdot0$ keV. Below 1·5 keV the flux is less than expected. This may be due to interstellar absorption. Giacconi (1966) has shown that interstellar absorption is the fundamental limitation on x-ray studies at long wavelengths. The degree of absorption depends critically on the density and chemical composition of interstellar space. Fig. 8.8 shows the distance in interstellar space that will result in the attenuation of an x-ray flux by a factor of 1/10 as a function of wavelength. The interstellar density is taken to be 1·0 atoms/cm³. For $\lambda < 6$ Å the whole Galaxy is transparent, but at $\lambda > 100$ Å sources must lie within 100 light-years, if they are not to be seriously attenuated.

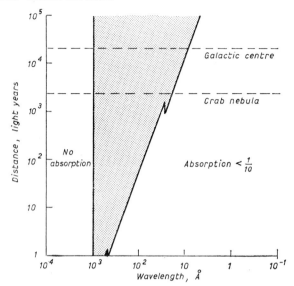

Fig. 8.8. Shaded region corresponds to interstellar absorption of x-ray flux by at least a factor of ten, where density $\sim$1 atom/cm^3

8.8 Variability

In late 1963 it became apparent that many of the x-ray sources were variable. This complicates their study and makes it difficult to combine the results of experiments made at different times. Since the different groups working in the field use different detection techniques and uncertain absolute calibrations, the conclusions about variations must be qualitative.

The first strong evidence for variations come from a comparison of two surveys of the Cygnus region by the N.R.L. group. In a survey in June 1964 CYG XR-1 was slightly stronger than CYG XR-2; in April 1965, CYG XR-1 had decreased so that it was only 0·25 times the intensity of CYG XR-2. In September 1964 the Lockheed group had estimated the intensity of CYG XR-1 as only one-sixth that of the N.R.L. June value. CYG XR-2 appeared to be unchanged over this period although in April 1965 balloon observations at higher energies failed to detect it, but did detect CYG XR-1. In March 1966 the position was reversed, with the flux from CYG XR-1 being less than 10% that of CYG XR-2.

The position in the Galactic centre is complex with large discrepancies between different surveys which can only be explained by fairly rapid time variations. Two of the strongest and most studied sources, SCO XR-1 and TAU XR-1, have not yet displayed any strong variations. This is surprising since the optical luminosity of SCO XR-1 is changing and there is evidence for continuing violent activity in the Crab Nebula.

Data on variations are still fragmentary and difficult to correlate. It is unlikely that this situation will change until x-ray experiments in satellites provide a means of continuous observations of one source with one detector over a long period.

8.9 X-ray background

In all the x-ray rocket and balloon experiments the sources appear to be superimposed on a background of x-rays whose origin is apparently extra-terrestrial. As the angular resolution and sensitivity of x-ray detectors has improved, more and more sources have been resolved from this background. Unlike the discrete source measurements in which relative anisotropies are of importance, background measurements must be absolute, that is, the detector must only be sensitive to extra-terrestrial x-rays. Since it is virtually impossible to build such a detector, all measurements on the diffuse background have a large degree of ambiguity. Although local particles can be immediately discounted as an explanation for discrete sources, the possibility that particles, in particular low-energy electrons, make a significant contribution to x-ray background measurements cannot be completely rejected. Since all experiments thus far have been performed with a few grams of atmosphere above the detectors, there is always the possibility that the counts recorded are secondary x-rays produced above the detector or albedo x-rays from the atmosphere below. The latter contribution can be corrected for by recording the background rate when the detector is looking towards and away from the earth. The difference gives the flux incident on the detector from the thin layer of atmosphere above it and interstellar and inter-galactic space.

In the 2 to 8 Å range the background flux is 10 photons/cm² sec sterad; this is believed to be of extra-terrestrial origin. This radiation appears to be isotropic; there is no evidence for an increased rate from the Galactic disc which would point to a Galactic origin. The most likely source is therefore extra-galactic space. It has also been suggested that most, if not all, of the background radiation is a composite of discrete sources and that improved experimental techniques will enable these to be resolved. Since most sources are Galactic the background would be expected to be stronger in the region of greater source concentration, that is, along the Galactic plane. The contribution to the background, assuming all normal galaxies have the same x-ray luminosity as the Galaxy, has been estimated by the N.R.L. group to be one-tenth the measured background. Evolutionary effects could enhance the background from these sources to the measured value.

The background measurements, even if only upper limits, are sufficient to eliminate one cosmological model of the universe. In one version of the steady-state theory it was proposed that matter was continuously created in the form of neutrons. The resulting decay-electrons would heat the intergalactic medium to a temperature of $10^9 \,^\circ$K. This would cause the emission of thermal bremsstrahlung x-ray photons; the flux is of the order of 50 photons/cm^2 sec sterad which is five times the observed flux.

8.10 Summary

The small number of identifications that have been made enable a tentative classification of x-ray sources to be made. The principal features of the two types are shown in Table 8.3.

TABLE 8.3

Classification of x-ray sources

TYPE	I	II
Examples	TAU XR-1 CYG XR-1	SCO XR-1 CYG XR-2
Spectrum	$I(\nu) \propto \nu^{-2}$	$I(\nu) \propto \nu^{-1}\, e^{-h\nu/kT}$
Mechanism	Magnetic bremsstrahlung	Thermal bremsstrahlung
Optical Object	Diffuse	Stellar

8.11 Source models

Not enough data has yet been accumulated to permit detailed x-ray source models to be drawn up. A number of proposals have been made but it is too early yet for anything other than qualitative discussions of the emission mechanisms involved. It is quite likely that the observed sources are examples of a number of different phenomenon, whose classification will only be possible when each source has been studied in as much detail as SCO XR-1. The principal proposals will be summarized below.

(A) NEUTRON STAR HYPOTHESIS

Neutron stars were proposed long before x-ray astronomy became a possibility. The ultimate fate of stars which undergo collapse to nuclear densities has always interested those who have derived the detailed models of stellar evolution. Because the objects are so highly collapsed, they are not detectable at radio or optical wavelengths. Chiu (1964) suggested that the high surface temperature of these objects (10^7 °K) would give detectable x-ray emission if the object radiated like a blackbody. This proposal was made primarily to explain the sources SCO XR-1 and TAU XR-1. These neutron stars could be the end phase of a supernova explosion. Morton (1964) extended Chiu's proposal and developed a model with the following parameters:

Total mass $\qquad M = 1\cdot3\ M_\odot$

Radius $\qquad R = 9\cdot25$ km

Average density $\qquad \bar\rho = 7\cdot85 \times 10^{14}$ g/cm^3

For this model Morton calculated the x-ray luminosity and lifetime for neutron stars with central temperatures ranging from 2×10^7 °K to 2×10^9 °K. He concluded that the observed sources could be explained by an object with a central temperature of about 5×10^8 °K and a surface temperature of about 8×10^6 °K.

Subsequent research has led to the rejection of the neutron star hypothesis of x-ray source origins. Theoretical studies have shown that if the energy loss by neutrino emission is taken into account, the high temperatures postulated for the centre of the neutron star cannot be sustained and the star will cool in about a year. Furthermore the lunar occultation experiment on TAU XR-1 by the N.R.L. group showed that the source had finite dimensions whereas a neutron star would appear like a point source at that distance.

(B) MAGNETIC BREMSSTRAHLUNG MODEL

The electromagnetic spectrum of continuum radiation from the Crab Nebula at radio and optical frequencies has already been discussed in Chapter III. The most satisfactory explanation of the observed power law spectrum seems to be in terms of the magnetic bremsstrahlung mechanism. If the optical power law $I(\nu) = k'\ \nu^{-1.1}$ is extrapolated, the resulting curve passes through the x-ray flux point from the source TAU XR-1. The size of this source is comparable with the size of the amorphous optical source on which it is centred. The spectrum steepens at x-ray frequencies and can be represented by a power law

$$I(\nu) \propto \nu^{-\alpha}$$

where

$$\alpha = 1.4, \qquad h\nu = 1.5 \text{ to } 6 \text{ keV}$$
$$\alpha = 2.0, \qquad h\nu = 20 \text{ to } 60 \text{ keV}.$$

This power law dependence suggests that the x-ray part of the spectrum may also be due to magnetic bremsstrahlung radiation. Since the source CYG XR-1 has a similar spectrum at high energies it may be another example of the same phenomenon.

This model has a number of important consequences, the most important of which is the presence of large numbers of ultra-relativistic electrons in the Crab and similar sources. From Chapter 2 the characteristic frequency of electrons of Lorentz factor γ in a field H is given by

$$v_s \sim 4.2 \times 10^{-6} \, \gamma^2 \, . \, H.$$

For $E = 40$ keV, $v_s = 10^{19}$ c/s; the possible values of γ and H are given in Table 8.4. At these high energies the dominant loss mechanism

TABLE 8.4

X-ray magnetic bremsstrahlung

H (gauss)	γ	E (eV)	$t_{\frac{1}{2}}$ (sec)
10^{-5}	4.9×10^8	2.5×10^{14}	1.6×10^{10}
10^{-4}	1.55×10^8	7.9×10^{13}	5.1×10^8
10^{-3}	4.9×10^7	2.5×10^{13}	1.6×10^7
			(1 year $\sim 3.2 \times 10^7$ sec)

is magnetic bremsstrahlung radiation, so that the time for an electron to lose half of its initial energy is given by equation (6.1).

For reasonable values of $H(\sim 10^{-3} - 10^{-4}$ gauss from equipartition arguments), the lifetime of the electrons is extremely short. Since the source lifetimes are far in excess of these electron lifetimes, e.g. the Crab Nebula is 900 years old, the supply of high-energy electrons must be constantly being replenished. This, in turn, suggests that there is a very efficient acceleration mechanism at work in the Crab to the present day. The supernovae remnant is then an active source, with a steady supply of relativistic particles, not merely the slowly decaying aftermath of a stellar explosion. This conclusion does not follow from the radio and optical observations, since for moderate magnetic fields the life-times of the electrons radiating at these wavelengths is greater than the age of the nebula. Hence the electrons now radiating at these wave-lengths could have been accelerated in the original outburst.

One method by which relativistic electrons could be continuously

K

produced is via the decay of charged mesons produced in proton-proton collisions. If a flux of relativistic protons were produced in the original explosion, then they would be retained in the source by the magnetic field and would only lose their energy slowly, primarily in collisions with the gas in the nebula. In these collisions π mesons would be produced and hence electrons and positrons of comparable energies. Some of the protons would leak into interstellar space and constitute a portion of the primary cosmic radiation.

If this mechanism were operative, then high-energy gamma-rays should also be produced via the decay of π° mesons which are also produced in the proton-proton collisions. Gould and Burbidge (1965) have calculated the gamma-ray flux expected from the Crab Nebula from this mechanism and conclude that the flux would be greater than the upper limits that have been set to the gamma-ray flux with energies greater than 10^{12} eV. Some other mechanism must be found for the continuing replenishment of the relativistic electron supply if the hard x-ray sources are to be fitted to this model.

The validity of this magnetic bremsstrahlung model will not be determined until x-ray polarization measurements have been made. The measurement of polarization in weak fluxes is difficult but a number of groups are developing experimental techniques which should settle this question in the near future.

(C) THERMAL BREMSSTRAHLUNG

The Coulomb scattering of electrons by nuclei in gaseous matter can give rise to photons in the x-ray region. This thermal or collisional bremsstrahlung can come from thermal electrons in a hot optically thin gas or from very hot electrons in a cold plasma. The former is the more important and has been proposed as the basic emission mechanism for the 'old nova' x-ray sources. If the electrons have a Maxwellian distribution, then the resulting spectrum is of the form

$$I(\nu) \propto \exp(-h\nu/kT).$$

This gives a soft x-ray spectrum with $T \sim 10^7$ °K.

The classification of the optical counterpart of SCO XR-1 as an old nova is not yet certain. Many old novae have properties similar to this object but are not x-ray sources. They are often surrounded by a halo of hot gas from which optical bremsstrahlung is observed. If the exponential x-ray spectrum of SCO XR-1 is produced into the visible, then much of the observed visible continuum can be associated with this process. Since this latter component is highly variable, equally rapid variations would be expected in the x-ray luminosity. It is worth noting that the energy emitted in the 1 to 10 Å region is 10^3 times that emitted in the visible. Although the object may be an old nova, energetically it is primarily an x-ray source. If this mechanism is operative and if normal chemical abundances are assumed, then x-ray line emission would be expected. The 1·02 keV line of Ne should be resolvable by instruments with only moderate energy resolution. Fine energy resolution should reveal many other lines which may ultimately lead to a detailed understanding of the processes taking place in the source.

Because of the harder spectrum from the supernovae remnant group of sources, this mechanism is less likely to be operative for them. The plasma temperature for TAU XR-1 would have to be at least 10^8 °K, which would require a continuous supply of thermal energy. A source of such energy has been proposed by Morrison and Sartori (1965). Heat is supplied by the radioactive decay of heavy elements produced in a supernova explosion. The exponential form of the light curve for Type I supernovae suggests that radioactive decay may be the dominant energy source. The production of large quantities of Californium 254, which has a half-life of 55 days could give the observed effect. The x-ray emission would then come from the hot dilute gas plasma surrounding the hot decaying nucleus.

9
Gamma-ray Astrophysics

9.1 Introduction

Like x-ray astronomy, gamma-ray astrophysics is still in its infancy; unlike x-ray astronomy these early years have not been characterized by any surprising results. That the wide variety of techniques, which must be used over the wide range of energies, have not been sensitive enough to isolate astronomical fluxes of gamma-rays does not mean that useful results have not been obtained; already upper limits have been set which eliminate predicted fluxes in certain models. Gamma-ray fluxes must most certainly exist; because of their high energy, they are closely related to many of the relativistic particle source systems and hence are of very great theoretical interest.

The importance of the study of gamma-rays as a means of probing the universe has been emphasized in a recent comprehensive review article by Fazio (1967). Gamma-rays are unique in that they are the only detectable high-energy quanta which can reach us from distant sources. Charged particles are deflected by magnetic fields whereas neutrons and other neutral particles will decay *en route*. Neutrinos will arrive with direction intact, but are extremely difficult to detect.

Fazio lists five fundamental problems of astrophysics which gamma-ray studies might elucidate:

(i) Cosmic ray origins.
(ii) Cosmic ray densities, both in the Galaxy and in inter-galactic space.
(iii) Density and composition of Galactic and inter-galactic matter.
(iv) Density of anti-matter in the universe.
(v) Strength of Galactic and inter-galactic magnetic fields.

The experimental techniques available in gamma-ray astronomy have been discussed by Greisen (1966). Several techniques are currently being revised to improve their sensitivity: it can be confidently predicted that gamma-ray fluxes will be recorded in the next decade, although at what energy or from what source is still a matter of speculation. If radio or x-ray astronomy is anything to go by, the strongest gamma-ray sources will not be the objects expected from studies in other parts of the electromagnetic spectrum.

9.2 Production mechanisms

The first concrete proposal of the existence of detectable astrophysical gamma-rays came in a review paper by Morrison in 1958. Although many of the production processes proposed have since been subjected to a more rigorous treatment, this paper contains the essence of the theoretical background to gamma-ray astrophysics. Since the photon production processes at very high energies are not as well understood as in, say, the x-ray region, the flux estimates are of necessity more uncertain. In general the uncertainty in the astrophysical parameters is sufficiently great to be the dominant factor in the uncertainty of these flux estimates.

The most important production processes, particularly in interstellar and intergalactic space, have been treated in detail in several reviews. Estimates have also been made of the fluxes expected from the most energetic astronomical objects, where the parameters must be chosen somewhat arbitrarily. Experimental techniques are more sensitive to gamma-ray fluxes from discrete sources because it is easier to detect an anisotropy than to make an absolute measurement.

(A) CONTINUUM PROCESSES

The following three processes are the most important contributors to the gamma-ray continuum using generally accepted values of astrophysical constants.

(i) *Neutral π-meson decay*. The π^0 meson decays to two gamma-rays with a lifetime of 10^{-16} seconds

$$\pi^0 \xrightarrow[10^{-16} \text{ sec}]{} \gamma + \gamma.$$

In the centre-of-mass system the energy of each gamma-ray is 70 MeV; in the LAB system the gamma-ray energy has a range of values centred on 70 MeV.

The most important source of π^0 mesons is creation in proton–proton collisions where the incident proton must have an energy greater than 290 MeV.

$$p + p \rightarrow p + p + n_1 \, (\pi^+ + \pi^-) + n_2 \, \pi^0$$
$$p + p \rightarrow p + n + \pi^+ + n_3 \, (\pi^+ + \pi^-) + n_4 \, \pi^0 \text{ etc.}$$

where n_1, n_2, n_3, n_4 are small whole numbers. This process will occur in interstellar space, where the interstellar gas (mainly hydrogen) is bombarded by the cosmic radiation (mainly protons). The cosmic-ray energy spectrum at the earth is well-known; the same spectrum may be assumed for the radiation in interstellar space. To a first approximation the radiation can be considered to be composed of just protons. The interstellar density is non-uniform and finite in extent. To calculate the gamma-ray flux from the source, the interstellar density must be integrated along the line of sight. Therefore, the effective number of target atoms,

$$\mathcal{N} = \int_0^L \rho(l)\, \mathrm{d}l$$

where $\rho(l)$ = interstellar density at a distance l, and L = distance to the edge of Galaxy along line of sight.

The shape of the gamma-ray energy spectrum is somewhat steeper than the incident proton spectrum, e.g. if

$$I(>E)_{\text{proton}} \propto E^{-1.6}$$

then
$$I(>E)_{\text{gamma}} \propto E^{-1.8}$$

There are a number of uncertainties in the cross-sections and multiplicities at high energies; empirical values may be taken from the relation of the secondary gamma-ray flux in the atmosphere to the primary cosmic ray spectrum. The flux of neutrinos and electrons from the decay of charged mesons can be calculated similarly.

In Table 9.1a the gamma-ray fluxes from the process, as estimated

TABLE 9.1

(a) *Galactic radiation* $I(> 50 \text{ MeV})$, *photons/cm² sec ster.*

MECHANISM	GALACTIC CENTRE	GALACTIC POLE
π^0 decay	2×10^{-4}	2×10^{-6}
Bremsstrahlung	4×10^{-4}	4×10^{-6}
Compton	5×10^{-5}	10^{-4}

(b) *Radiation from extra-galactic space*
$I(> 50 \text{ MeV})$ *photons/cm² sec ster.*

MECHANISM	
π^0 decay	$4 \times 10^{-4}\, k_{cr}$
Bremsstrahlung	$6 \times 10^{-4}\, k_e$
Compton	$4 \times 10^{-2}\, k_e$

k_{cr} = ratio of inter-galactic cosmic ray spectrum to Galactic spectrum.
$_e$ = ratio of inter-galactic electron spectrum to Galactic spectrum.

by Ginzburg and Syrovatskii (1964), are given for energies greater than 50 MeV. Two sources are considered: interstellar space looking towards the Galactic centre and the Galactic pole. Table 9.1*b* gives the predicted flux from extra-galactic space. The difference in the fluxes from the two Galactic sources, if detected experimentally, would confirm their origin and be an important confirmation of the parameters used in the calculation.

π^0 mesons are also produced in proton-anti-proton annihilations. The cross-section for annihilation is inversely proportional to the relative velocities of the two particles. This process is important primarily at low energies so that a sharp peak would be expected at 70 MeV. A large flux from this process is dependent on anti-matter being a major constituent of the universe.

At very high energies π^0 mesons will be produced in proton-photon collisions. The threshold proton energy for π^0 production in a head-on collision between a proton and photon is given by:

$$E_{\text{proton}} > \frac{m_\pi}{2e}(m_\pi + 2M)$$

where m_π = mass of π meson, M = mass of proton, and e = photon energy.

The average energy of the photons associated with thermal stellar radiation is about 2 eV. This process is therefore important for $E_{\text{proton}} > 10^{17}$ eV. The resulting gamma-rays will have energies between 10^{15} and 10^{16} eV. Estimates of the gamma-ray flux from this process show that it is 10^{-5} times the cosmic-ray proton flux at the same energies. Microwave photons, by the same process, will cause a serious energy loss for protons with $E > 10^{20}$ eV (Chapter 11), and will produce gamma-rays with energy between 10^{18} and 10^{19} eV. This gamma-ray flux will also be only a small fraction of the charged particle flux at these energies.

(*ii*) *Bremsstrahlung*. The deceleration of relativistic electrons in the Coulomb fields of charged particles produces photons with energies of the same order as the electron energy. To estimate the gamma-ray flux expected by this process it is necessary to know the electron energy spectrum and the charged particle density in the region of interest. The probability of the bremsstrahlung production of a photon of energy E_γ by an electron of energy E is given by:

$$P(E_\gamma, E)\, dE_\gamma = \frac{dE_\gamma}{E_\gamma} \frac{x}{X}$$

where x = thickness of matter traversed, and X = radiation length (in interstellar space $X \sim 60$ g/cm^2).

The gamma-ray fluxes, estimated by Ginzburg and Syrovatskii, from this process are given in Table 9.1.

(*iii*) *Compton.* Gamma-rays may also be produced by the Compton effect, the physics of which has been treated in Chapter 2. The form of the gamma-ray spectrum was shown to depend on the relativistic energy spectrum and the photon energy density. A thermal distribution of the latter is usually assumed.

Felten and Morrison have calculated the x-ray and gamma-ray flux expected by this process from (1) the Galactic halo (2) extra-galactic space. In each case they find that the more intense flux arises from the collision of electrons with the 3 °K black-body microwave field (see Chapter 11). Only for gamma-rays with energies greater than 100 MeV does the thermal spectrum from stellar sources in the visible region play a more important role. Using the electron energy spectrum discussed for the Galactic halo in Chapter 4, the gamma-ray energy spectrum has the form shown in Fig. 9.1, where the low energy cut-off

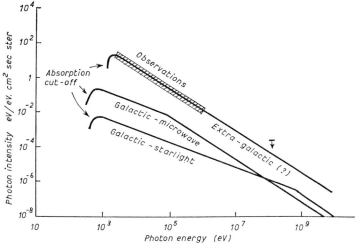

Fig. 9.1. Photon energy spectrum from Compton effect

is due to x-ray absorption by the interstellar gas, and the knee corresponds to a break in the primary electron spectrum at $E \sim 10$ GeV.

To calculate the gamma-ray contribution from extra-galactic space, it is necessary to assume the electron energy spectrum. At the present time this can only be arrived at by somewhat arbitrary methods. One approach is to assume that all galaxies have a halo similar to the Galaxy and hence deduce the extra-galactic contribution from the super-position of all such sources. This flux is an order of magnitude lower than that from the halo of the Galaxy. Some galaxies, e.g. radio galaxies, may be much stronger emitters than the average galaxy so that the total flux may be greater. Also the flux in a cluster of galaxies such as the Local Cluster or the Virgo Cluster may be greater than the general extra-galactic flux. The fluxes estimated by Ginzburg and Syrovatskii are given in Table 9.1.

(B) LINE RADIATION

Although the continuous spectrum of gamma-rays probably constitutes the greater proportion of cosmic gamma-rays, the few processes that give distinct line emissions may be ultimately more important for the understanding of the conditions in astrophysical situations. The most important of these, which give gamma-ray lines at low energies, i.e. less than 10 MeV, will be briefly considered.

(i) *Electron–positron annihilation.* The electron–positron annihilation reaction

$$e^+ + e^- \longrightarrow \gamma + \gamma$$

will give gamma-rays whose energies are strongly peaked at the rest mass of an electron, i.e. 0·51 MeV. The fluxes to be expected depend on the density of positrons. One of the most important positron production processes is through the decay of positively charged mesons. The cross-section for annihilation of a positron of energy E is given by:

$$\sigma(E) = \pi r_0^2 \frac{mc^2}{E} \ln \left(\frac{2E}{mc^2} - 1 \right)$$

where r_0 = classical electron radius. The probability of annihilation depends on the cross-sections of other interactions which would absorb the positron before it has been slowed sufficiently for the annihilation cross-section to be appreciable. The detection of this line, both from the diffuse background and from discrete sources, will be one of the most important contributions of the study of gamma-rays to astrophysics.

(ii) *Nuclear transitions.* A number of nuclear processes could give gamma-rays of detectable intensity. The reaction

$$p + n \longrightarrow d + \gamma$$

is expected when MeV neutrons are released by nuclear interactions into regions of density in excess of $10^{-9}\text{g}/\text{cm}^3$. This results in a flux of 2·23 MeV gamma-rays. Also cosmic-ray collisions with gas molecules can lead to the excitation of the nuclei with emission of characteristic gamma-rays. Flux estimates depend on the chemical composition assumed for the region of interest. Gamma-rays from the decay of radioactive materials are expected from discrete sources, such as supernovae, where heavy elements might be produced in large quantities. Of particular interest is the 0·39 MeV line from Californium[249].

In general the most optimistic estimates of the fluxes from any of these line processes are lower than the present sensitivity of gamma-ray detectors which have poor energy resolution.

(c) DISCRETE SOURCES

The experimental detection of gamma-rays from discrete sources is technically more feasible than the identification of the diffuse background. For this reason particular attention has been given to those sources from which detectable fluxes might be expected. The most detailed calculations have been for the Crab Nebula; the principal predictions for this source have been listed by Fazio and are shown in Table 9.2. The wide spread in the predicted values is a measure of the uncertainty of the parameters even in this much studied source. The

TABLE 9.2

Predicted fluxes from the Crab Nebula

ENERGY	MECHANISM	FLUX	REFERENCE
30 keV → 2 MeV	N.D.	10^{-2}	M
30 keV → 2 MeV	N.D.	10^{-2}	S
30 keV → 2 MeV	N.D.	10^{-4}	CC
100 keV → 1 MeV	B	6×10^{-5}	S
>1 MeV	C.S.	2×10^{-12}	S
>10 MeV	B	2×10^{-7}	S
>100 MeV	P.D.	$1·5 \times 10^{-5}$	GK
>100 MeV	B	3×10^{-6}	GK
>100 MeV	C.S.	7×10^{-8}	GK
>100 MeV	P.D.	10^{-4}	H
>100 MeV	C.S.	10^{-6}	G
>100 MeV	P.D.	2×10^{-9}	GS
>100 MeV	B	5×10^{-10}	GS
>100 MeV	C.S.	2×10^{-9}	GS
$>5 \times 10^5$ MeV	C.S.	4×10^{-10}	G
$>5 \times 10^6$ MeV	C.S.	10^{-12}	G
5×10^6 MeV	P.D.	8×10^{-9}	GB

N.D. = nuclear decay
B = bremsstrahlung
C.S. = Compton scattering
P.D. = π^0 decay

M = Morrison (1958)
S = Savedoff (1959)
CC = Clayton and Craddock (1965)
GK = Garmire and Kraushaar (1965)
H = Hayakawa *et al.* (1964)
G = Gould (1965)
GS = Ginzburg and Syrovatskii (1964)
GB = Gould and Burbidge (1965)

fluxes predicted at high energies are particularly interesting because of their relevance to cosmic ray origin and acceleration problems.

If the visible light from strong radio sources is magnetic bremsstrahlung radiation from relativistic electrons which are the decay products of protons produced in proton–proton collisions, then Cocconi (1960) predicted that a flux of gamma-rays with energies greater than 10^{12} eV should be produced which would be detectable with cosmic-ray air shower techniques at mountain altitudes. Equal numbers of gamma-rays and electrons are produced so that if a magnetic field of 10^{-4} gauss is assumed, the flux expected is given by

$$I(> E)_{gamma} = 10^{(6 \cdot 7 - m/2 \cdot 5)} \, \gamma^{-1} \text{ photons/m}^2/\text{sec}$$

where $\gamma =$ Lorentz factor of electrons of energy E and $m =$ visual magnitude.

For the Crab Nebula this gives a flux:

$$I(>10^{12} \text{ eV})_{gamma} = 3 \times 10^{-7} \text{ photons/cm}^2/\text{sec}$$

which is well above the nuclear background and easily detectable.

Gould and Burbidge (1965) have considered the same process in the Crab Nebula in more detail. They consider two cases: (a) The spectrum from radio and optical frequencies is magnetic bremsstrahlung in origin. In this case the gamma-ray flux is

$$I(>100 \text{ MeV})_{gamma} = 5 \cdot 7 \times 10^{-4} \text{ photons/cm}^2/\text{sec}.$$

(b) The entire spectrum from radio to x-ray frequencies is magnetic bremsstrahlung radiation. The most significant flux is then at the higher energies:

$$I(>5 \times 10^{12} \text{ eV})_{gamma} = 8 \times 10^{-9} \text{ photons/cm}^2/\text{sec}.$$

Ginzburg and Syrovatskii have calculated the gamma-ray fluxes from a number of strong radio sources for the most important radiation mechanisms: π^0 decay, bremsstrahlung and Compton. These fluxes are listed for the Crab Nebula in Table 9.2. The same authors have calculated the flux expected by the Compton mechanism in quasars. Since the dimensions of these objects relative to their brightness is small, this mechanism is probably the most significant. For 3C 273 the electron injection power must be 10^{47} ergs/sec and the loss due to Compton scattering 5×10^{45} ergs/sec. For a radius of 4×10^{16} cm this would give a gamma-ray flux:

$$I(>10 \text{ MeV})_{gamma} \sim 10^{-5} \text{ photons/cm}^2/\text{sec}.$$

If a larger radius is taken, the gamma-ray flux is lowered by a corresponding amount.

The Compton effect in the Crab Nebula has been considered in detail by Gould (1965). This is one of the few sources for which reasonably reliable estimates of the parameters can be made. In this model the collisions between the relativistic electrons and the magnetic bremsstrahlung photons are considered. The resulting power law gamma-ray spectrum has an exponent similar to the magnetic bremsstrahlung spectrum. At the highest energies the Klein–Nishina cross-section must be used, giving a steeper spectrum. The complete integral

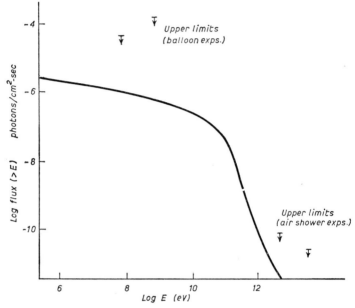

Fig. 9.2. Integral gamma-ray flux predicted from Crab Nebula from Compton effect

gamma-ray spectrum for model (*b*) of Gould and Burbidge (above) is shown in Fig. 9.2 for a magnetic field of 10^{-4} gauss. The fluxes are:

$$I(>100 \text{ MeV})_{\text{gamma}} = 10^{-6} \text{ photons/cm}^2/\text{sec}$$
$$I(>5 \times 10^{11} \text{ eV})_{\text{gamma}} = 4 \times 10^{-10} \text{ photons/cm}^2/\text{sec.}$$

9.3 Experimental techniques

(A) LOW ENERGY RANGE (<30 MeV)
The nuclear transition region, which is one of the most interesting regions of the cosmic gamma-ray spectrum from a theoretical standpoint, is one of the most difficult regions to explore experimentally [Greisen (1966)]. It is unlikely that significant results can be achieved with the techniques at present available; an increase of at least an

order of magnitude in sensitivity is required before really interesting results will materialize.

The simplest detector in this energy region is the so-called Phoswich detector, which utilizes the response of different scintillator materials to photons and charged particles. An alkali halide crystal, e.g. caesium iodide, is particularly sensitive to photons; on the other hand a thin plastic scintillator is virtually transparent to photons but detects charged particles with high efficiency. The Phoswich detector consists of an alkali halide crystal surrounded by a plastic scintillator. Both materials are viewed by the one photomultiplier; because the scintillation times, and hence the photomultiplier pulses, from the two materials are quite different, they can be easily sorted electronically. A gamma-ray event is characterized by an output pulse from the alkali halide crystal but not from the plastic. If the alkali halide crystal is thick enough for the secondary electron to be absorbed, then the light output will be proportional to the gamma-ray energy. Energy resolution of the order of 10% can be achieved with this device. Since this is the region where lines are expected, energy resolution is of paramount importance. It is necessary to calibrate these detectors with gamma-rays of known energy. It is estimated that a cubic crystal of side 3 inches, if exposed in a satellite for a year, could distinguish a line whose intensity was one-thousandth that of the diffuse background.

Since these detectors have no directional properties, they are not suitable for the investigation of discrete sources. The Compton scattering process, which is the dominant absorption process at these energies, is difficult to utilize for directional measurements. The cross-section for pair production is low and the opening angle of the pair large. Collimation is therefore required as in x-ray detectors. A passive collimator, consisting of a lead honeycomb, can weigh up to 500 kilograms since the absorption length for MeV gamma-rays in lead is of the order of an inch and the sides and rear of the detector must be shielded. An active collimator consists of an anti-coincidence scintillation counter shield with a window as the effective aperture. However, the sensitive area must be small if the effective running time is not to be reduced to a negligible amount by the high rate of anti-coincidence veto signals.

(B) MEDIUM ENERGY RANGE (30 MeV–1 GeV)

At energies above 30 MeV, gamma-ray detection techniques invariably use the pair production reaction; at energies below 1 GeV detectors can be made sufficiently large for the reaction and the absorption of the product to take place within the detector.

The probability that a photon of energy E produces an electron pair per g cm^{-2} of absorber is

$$P(E) = k\frac{Z^2}{A}E\left\{\frac{7}{9}\ln\left(\frac{183}{Z^{1/3}} - \frac{1}{54}\right)\right\}$$

where k is a constant and Z and A refer to the absorber. A pair-production event has a definite signature which enables it to be identified amongst a number of spurious events. The characteristics of the pair-production event, and the way in which they are utilized, are listed below:

(i) Z *dependence.* The relative transparency of plastic scintillation materials of low Z makes them particularly suitable for anti-coincidence shields. A lead converter is usually employed in which the probability of pair production is high. In some instruments this conversion layer is omitted and pair production takes place in the detector itself, e.g. in the metal plates of a spark chamber or with the silver atoms in a nuclear emulsion. This has the advantage that the electron pair is registered immediately after creation before multiple scattering distorts the trajectories.

(ii) *Identification of event.* To identify the pair-production event some visual means of recording the electron-positron tracks must be utilized. Two tracks with small divergence angle appearing well within the detector, with no trace of a parent particle, can usually be regarded as a gamma-ray event. Spark chambers or nuclear emulsions are the most useful methods of visual recording. The angular resolution of the latter is lost due to the long time exposure necessary to record a reasonable number of events.

(iii) *Angular resolution.* The opening angle of the pair is of the order of mc^2/E radians where E is the gamma-ray energy. Angular resolution of the order of a few degrees can be obtained for $E > 100$ MeV. Nuclear emulsions, in combination with spark chambers, have the greatest promise for the precise positioning of a source on the celestial sphere.

(iv) *Energy measurement.* All of the energy of the gamma-ray is transferred to the electron pair. A detector which is sufficiently large that the electron pair is absorbed can in theory measure the energy of the gamma-ray. In practice, energy measurements in this region have not very much significance since no lines are expected; however, the general shape of the gamma-ray spectrum is of importance and most instruments make a crude measurement of the gamma-ray energies.

A typical experiment (using a spark chamber, scintillation detectors and a Cherenkov detector) designed to detect cosmic gamma-rays at balloon altitudes is shown in Fig. 9.3. This type of experiment is

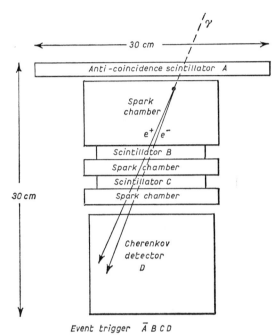

Fig. 9.3. Detector arrangement for balloon gamma-ray experiment

particularly suitable for the investigation of discrete sources, since at balloon altitudes the diffuse cosmic gamma-ray flux will be masked by the large flux of secondary gamma-rays, the products of the inter-actions of cosmic rays with the upper layers of the atmosphere. The diffuse cosmic gamma-ray flux is estimated as less than one-tenth the secondary flux at a depth of 5 g/cm².

An alternative experimental arrangement is shown in Fig. 9.4 in which a combination of scintillation and Cherenkov detectors are used to define a gamma-ray event without the use of a visual detector. This arrangement was used in the Explorer XI satellite; the complete detector weighed only 30 lb and had a field of view of 40°, defined by the geometry of the counters. The efficiency of the detector for gamma-rays was 0·15. Although the background is much less at satellite altitudes, it is not possible to dismiss it altogether: because this instru-ment had no visual display, the unique character of the pair-production event could not be fully utilized. Some contribution from particle

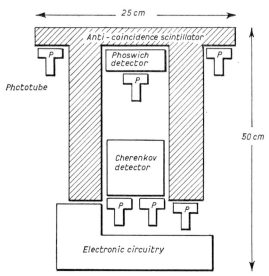

Fig. 9.4. Essential features of the Explorer XI gamma-ray detector

events cannot be completely ignored, so that the instrument is best suited for setting an upper limit to the diffuse background.

(c) HIGH-ENERGY REGION (>1 GeV)

The interval from 1 to 100 GeV is extremely difficult to investigate experimentally; because the fluxes expected are low, detectors with large areas are required. At present the weight of detector required is too heavy for exposure above the atmosphere by balloon or rocket. It is unlikely that there will be any significant development in this region until space technology has solved the problem of putting very massive objects in orbit.

At energies above 100 GeV, the gamma-ray has sufficient energy to cause an electromagnetic cascade in the atmosphere of dimensions sufficient for detection by extensive air shower arrays at mountain altitudes. The photon-initiated air shower has many features similar to the proton-initiated shower but lacks the strong nuclear core. Although the development of the shower is a complicated statistical process, it is sometimes possible to identify the primary from a detailed study of the shower at detector level.

At energies above 10^{15} eV photon showers can be distinguished by the abnormally small number of μ-mesons at mountain altitude. This interpretation is not unambiguous; a distinct class of showers could arise from some peculiarity of high-energy nuclear interactions in the first stages of the showers. The Bolivian Joint Air Shower Experiment on Mount Chacaltya (Chapter 3) has 60 m² of shielded scintillation detectors, which, in association with unshielded detectors, enable the ratio of μ-mesons to electrons in the shower to be estimated. The direction of the primary can be fixed to within 4° using fast timing techniques. This large array may be used to measure the diffuse background of gamma-rays with energies above 10^{15} eV and also to search for discrete sources. At energies below 10^{15} eV the number of particles in the shower is not sufficient to identify the nature of the primary.

Gamma-rays from discrete sources will have one characteristic which can enable them to be distinguished from the cosmic-ray background: directional anisotropy. Unlike charged particles, gamma-rays retain their original directions and are not influenced by interstellar magnetic fields. Low-energy showers (10^{11} to 10^{15} eV) can be most conveniently detected by the Cherenkov light that the shower particles emit in their passage through the atmosphere. Unlike the particle component this light is not absorbed and spreads over an area of the order of 10^5 m² so that a small light detector can have a large collection area. An angular resolution of about 2° can be obtained. Since the light pulses from photon- and proton-initiated showers are similar, gamma-ray sources will only be apparent by the increase in shower counting rate when the source is in the field of view of the instrument.

9.4 Observations

(A) LOW ENERGY

Balloon observations are of little value in this region of the spectrum due to the large flux of secondary gamma-rays. These are the products of bremsstrahlung by relativistic electrons produced in cosmic-ray collisions in the upper atmosphere. Although early experiments sought to correct for the albedo, significant results had to await satellite experiments. At a depth of 14 g/cm² the upward albedo flux is one-half the downward flux. The greatest contribution of the early balloon experiments was that they led to the development of the Phoswich detector.

The first results of astrophysical importance came from two of the Ranger moon-rocket experiments [Arnold et al. (1962)]. Data were recorded on these in the region beyond the earth's radiation belts so that the earth's contribution could be neglected. Since the spacecraft itself could give rise to secondary gamma-rays, observations were made with the Phoswich detector in two positions; in a stowed position and at the end of a 6-ft arm. From the ratio of the counting rates in the

L

two positions, the importance of local secondaries could be estimated. The detector was almost omni-directional, but had good energy discrimination over the range 60 keV to 1 MeV. The spectrum obtained is plotted in Fig. 9.5. The differential spectrum may be represented by

$$N(E) \, dE = 0 \cdot 012 \, E^{-2 \cdot 2} \, dE \text{ photons/cm}^2 \text{ sec ster}$$

where E is in MeV. In an earlier satellite experiment, OSO-1, the

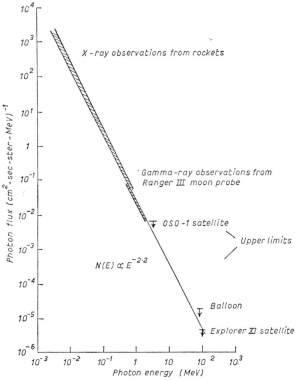

Fig. 9.5. Diffuse cosmic gamma-ray spectrum

results were suspect due to secondary contamination. The Ranger flights gave upper limits for the intensities of the two lines 0·51 MeV and 2·23 MeV of 0·0011 and 0·0004 photons/cm² sec ster. No detector has yet been flown with angular resolution suitable for the detection of discrete sources.

(B) MEDIUM ENERGY

In this energy region results have been principally obtained in balloon experiments. Using combinations of scintillation, Cherenkov and spark chamber detectors, angular resolution has been sufficient to place meaningful upper limits to the gamma-ray flux from discrete sources. Typical results are shown below:

Crab Nebula $I(> 100 \text{ MeV}) < 3 \times 10^{-5}$ photons/cm^2 sec

Cygnus A $I(> 1 \text{ GeV}) < 6 \times 10^{-4}$ photons/cm^2 sec.

In one experiment Duthie and his collaborators reported a statistically significant flux from the direction of Cygnus for gamma-ray energies greater than 100 MeV. A later flight by Frye and Wang, using an instrument with greater sensitivity, failed to confirm this result. All of these experiments were carried out in the presence of secondary background gamma-rays. It was not possible to estimate the diffuse cosmic background. Although the upper limits obtained are useful, a sensitivity of the order of 10^{-7} photons/cm^2/sec for $E > 100$ MeV is probably required if the fluxes predicted are to be detected. It is unlikely that present techniques can be extended by this amount.

As in the low-energy region, the future for this region seems to lie with satellite experiments. The very simple experiment of Kraushaar and Clark on Explorer XI has provided more astrophysical information than all of the balloon experiments. Because the instrument was free of the atmosphere, it obtained the best upper limits to the diffuse background. The duration of the flight was seven months, but since part of this time was spent in the earth's radiation belts or beyond the range of the receiving stations, only a fraction of the time was utilized. Since the identification of gamma-ray events was not unambiguous, the results were quoted as upper limits

All directions $I(> 100 \text{ MeV}) < (3 \cdot 3 \pm 1 \cdot 2) \times 10^{-4}$ photons/cm^2 sec ster.

Upper limits were also assigned to the fluxes from a number of discrete sources, but due to the poor statistics, these were inferior to those obtained with balloons. No anisotropy in the diffuse background was detected which might have indicated a Galactic origin. A very small increase in sensitivity could provide some very interesting results, since the theoretical predictions of the Galactic gamma-ray flux are considered quite reliable.

(c) HIGH ENERGY

No results on the diffuse background have been obtained from 10^9 to 10^{15} eV although some limits can be placed on the gamma-ray flux by indirect extrapolation of cosmic-ray experiments. At energies greater than 5×10^{12} eV, upper limits have been set to the fluxes from discrete sources by two groups using the atmospheric Cherenkov technique. A number of strong radio sources have been investigated, but although the sensitivity exceeded that of some predictions, no sources have been detected. Some of the limits obtained are listed in Table 9.3.

TABLE 9.3

Upper limits from discrete sources

SOURCE	ENERGY (eV)	UPPER LIMIT (photons/cm² sec)	REFERENCE
Crab Nebula	5×10^{12}	5×10^{-11}	Chudakov *et al.* (1961)
Virgo A	5×10^{12}	5×10^{-11}	
Cygnus A	5×10^{12}	5×10^{-11}	
3C 147	1×10^{13}	3×10^{-11}	Long *et al.* (1964)
3C 273	5×10^{13}	1×10^{-11}	

Above 10^{15} eV, results have come exclusively from two groups using large air shower arrays. The groups use different criteria for defining a possible gamma-ray shower and neither group claims to have positively identified gamma-ray showers; hence their results are upper limits. The characteristics of the two experiments are listed in Table 9.4. At energies greater than 10^{16} eV there is evidence for a fall-off in the relative number of mu-less showers.

TABLE 9.4

Diffuse background

GROUP	DETECTORS	$I(>10^{15}$ eV) (photons/cm² sec ster)
Polish–French [Firkowski *et al.* (1965)]	Shielded and unshielded Geiger counters	10^{-12}
BASJE [Hasegawa *et al.* (1965)]	Shielded and unshielded scintillator counters	10^{-13}

The BASJE experiment also features good angular resolution; upper limits to gamma-rays from discrete sources of 10^{-14} photons/cm² sec are indicated.

9.5 Absorption

The absorption of photons by the reaction $\gamma + \gamma \rightarrow e^- + e^+$ is important in high-energy gamma-ray astrophysics. This was first pointed out by Nikishov (1962) who considered absorption in interstellar space by optical photons. The cross-section for this process has the form

$$\sigma(E_1,E_2) = \frac{1}{2} \pi r_0^2 (1 - \beta^2) \left[(3 - \beta^4) \ln \left(\frac{1 + \beta}{1 - \beta} \right) - 2\beta(2 - \beta^2) \right]$$

where r_0 = classical electron radius, and βc = electron velocity in the centre-of-mass system.

The probability of absorption of a photon oₗ energy E_1 per unit path length is given by

$$P(E_1) = \sigma(E_1, E_2) \mathcal{N}(E_2) \, dE_2$$

where $\mathcal{N}(E_2)$ is the spectral density of low-energy photons. The threshold condition for this reaction is that $E_1 E_2 > (mc^2)^2$.

Nikishov assumed that the starlight spectrum $\mathcal{N}(E_2)$ was similar in form to that of the sun; he took the thermal photon energy density as 0·1 eV/cm³. Because of the threshold condition, absorption is serious only for gamma-rays in the 10^{11} to 10^{12} eV region.

If d = distance to the source and I_0 the intensity of the high-energy flux at the source, then

$$I(E_1) \text{ at earth} = I_0(E_1) \, e^{[-P(E_1) \cdot d]}$$

For $d = 10^{26}$ cm (characteristic distance of extra-galactic object), the absorption is serious.

The value of starlight density assumed by Nikishov is two orders of magnitude greater than currently accepted values of the inter-galactic starlight density. A re-evaluation of the starlight spectrum by Gould and Schreder considers the spectrum as composed of two parts (i) thermal spectrum from cool stars in nuclei of galaxies (ii) thermal spectrum from hot, young stars in the outer regions of spiral galaxies. These values are plotted in Fig. 9.6; also plotted is the absorption

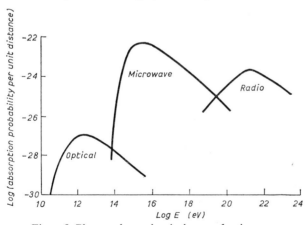

Fig. 9.6. Photon absorption in inter-galactic space

expected from the infrared spectrum resulting from fine structure level transitions in N⁺ ions in interstellar space. The absorption from these fluxes is relatively unimportant except for cosmological distances.

Goldreich and Morrison (1964) extended the work of Nikishov to take account of absorption of very high energy gamma-rays by radio photons. This radio background is rather difficult to estimate; these authors take 25% of the detected isotropic background flux as of extra-galactic origin. With this flux they find serious absorption at very high energies ($E_1 > 10^{18}$ eV). A revised estimate of the extra-galactic radio flux by Gould and Schreder gives $P(10^{20}$ eV$) \sim 10^{-25}$ cm^{-1}; since no techniques are available for detecting gamma-rays at these very high energies, this absorption is relatively unimportant.

The importance of the 3 °K field as a gamma-ray absorber was pointed out by Gould and Schreder (1966), and Jelley (1966). Assuming a thermal spectrum, the region of absorption is in the range 10^{14} to 10^{18} eV. In this region the absorption is some orders of magnitude greater than in other gamma-ray energy ranges because of the high number density of the microwave flux. The peak in the absorption occurs at 10^{15} eV, a region of particular interest in gamma-ray studies using extensive air-shower techniques. This absorption, taken in conjunction with the absorption at higher energy by radio photons, corresponds to complete absorption of gamma-rays with $E > 10^{14}$ eV which originate in extra-galactic sources. At the peak of the absorption curve $P(10^{15}$ eV$) \sim 10^{-22}$ cm^{-1} so that even gamma-rays from the Galactic centre $(d \sim 3 \times 10^{22}$ cm$)$ will be attenuated. The Crab Nebula $(d \sim 3 \times 10^{21}$ cm$)$ is not seriously obscured.

At lower energies some absorption by starlight is expected in the 10^{12} to 10^{13} eV region, but below 10^{12} eV there should be no absorption until the x-ray region is reached (at $E \sim 1$ keV photoelectric absorption becomes important). There is some observational evidence from air-shower studies of a break in the isotropic background gamma-ray spectrum at 10^{14} eV. The low absorption window at 10^{10} to 10^{13} eV is difficult to utilize as the fluxes are too low to be detected by balloons and the energies too low for the showers to be analysed. Refinement of high-energy gamma-ray techniques may permit the extra-galactic nature of the microwave flux to be confirmed.

9.6 Discussion

One of the most important aspects of gamma-ray studies is that the results may permit the determination, or at least the placing of limits, on some of the most important astrophysical parameters, which at the moment are either wholly unknown or inferred indirectly. Although most of the results to date have been in the form of upper limits, these upper limits have permitted the exclusion of several previously plausible theories. These results must be considered as only preliminary; the full value of gamma-ray studies has yet to be realized. Some of the conclusions that can be drawn from the existing rather scanty studies are listed below.

(A) DIFFUSE BACKGROUND

The flux predicted by Felten and Morrison (1966) from the collision of relativistic electrons with the 3 °K microwave field in the Galactic halo is two orders of magnitude lower than the observed diffuse background. In the region where the spectrum is dominated by Compton and magnetic bremsstrahlung loss by the electrons, the gamma-ray spectrum has the same slope as the observed spectrum. Assuming that this is more than coincidence, Felten and Morrison have extended their calculations. The possibility that the observed flux might be the sum of radiation from galaxies similar to the Galaxy has already been mentioned. Another possibility is that the flux might originate in inter-galactic space, again involving the 3 °K field and relativistic electrons. By fitting Compton gamma-ray spectrum to the observed results, the inter-galactic electron spectrum can be predicted. Over the entire range of interest Compton losses will predominate so the gamma-ray spectrum will have an exponent of -3.4 (Fig. 9.1). Below 10 keV the observed flattening may be explained by intergalactic absorption.

The electron intensity is 10^2–10^3 times less than the Galactic halo intensity. This is larger than is usually supposed to exist in inter-galactic space; if the extra-galactic theory of cosmic-ray origins is accepted, then these could result from collisions of the protons with the inter-galactic gas. Alternatively radio galaxies might be more efficient producers of electrons than is usually assumed. The observed clustering of the exponent of the radio flux, α, about 0·7, is evidence that the inter-galactic electron injection spectrum has an exponent $m = 2\alpha + 1 = 2.4$. If this flux of electrons is accepted, then the inter-galactic magnetic field must be less than 10^{-7} gauss; otherwise the magnetic bremsstrahlung background radiation would be detectable.

(B) SECONDARY PRODUCTION OF ELECTRONS

The upper limits obtained using the Cherenkov night-sky technique at 5×10^{12} eV are sufficiently low to disprove the Cocconi model for secondary production of the optically-radiating electrons in the Crab Nebula and other sources. These results are also applicable to the more refined model of Gould and Burbidge (1967). The upper limits are not sufficient to rule out the flux predicted by Gould from the Compton effect.

(c) MATTER-ANTI-MATTER ANNIHILATION

An upper limit can be placed on the amount of matter created, as required by the steady-state theory, if it is assumed that comparable amounts of matter and anti-matter are created in the same region. The upper limits to the gamma-ray flux set by the Explorer XI experiment indicate that the ratio of anti-protons to protons in inter-galactic space is $<10^{-6}$. Other satellite experiments give the same upper limit for the positron-electron ratio in inter-galactic space.

(d) INTER-GALACTIC ELECTRON DENSITY

From the Explorer XI results the inter-galactic electron intensity can be assigned a value at least ten times less than the intensity assumed on reliable grounds for the Galaxy.

(e) COMPARISON WITH RADIO AND X-RAY RESULTS

In view of the rapid development of the two other 'new astronomies' of this century, radio and x-ray, the rather slow progress of gamma-ray astronomy, particularly at lower energies, requires some explanation. Greisen (1966) has listed some of the difficulties associated with the detection of gamma-rays, in contrast to observations in other parts of the spectrum. Both in x-ray and radio astronomy the first measurements were made almost by accident and were surprising in their intensity. By contrast the early gamma-ray results have shown that the fluxes are somewhat lower than predicted in some optimistic models. The brightness of the discrete sources relative to the background in the two other disciplines made their detection feasible with instruments of comparatively poor angular resolution. It is obvious that gamma-ray sources will not be localized until the existing techniques can achieve resolutions about an order of magnitude greater than at present.

In gamma-ray measurements, as in x-ray, the sensitivity of the instruments is limited by the background particle radiation at balloon and rocket altitudes. At energies of 1 keV the cosmic-ray background is some 10^2 times lower than the diffuse x-ray flux; at energies of 1 GeV the gamma-ray flux is less than 10^{-3} the cosmic-ray background at the same energy. In contrast the photon flux at 1 keV is more than 10^6 times that at 1 GeV.

Balloon techniques do not permit loads of more than 2000 lb to be carried. This makes collimation of gamma-ray detectors prohibitive since the absorption length is ~ 10 g/cm^2 whereas that of x-rays is $\sim 10^{-3}$ g/cm^2. Secondary production of radiation both in the atmosphere and in the instrument shielding is also much more of a problem at gamma-ray energies.

10
Neutrino Astrophysics

10.1 Introduction

As with gamma-ray astrophysics, neutrino astrophysics is still speculative. Because the problems involved in neutrino detection mostly arise from its physics, the principal exponents are nuclear physicists, rather than astronomers. To date, no neutrinos of extra-terrestrial origin have been positively identified; there are, however, a number of experiments in course of construction capable of detecting the predicted fluxes from the sun. For sources other than the sun the predictions are not yet sufficiently precise to estimate the feasibility of their detection; the failure to detect gamma-ray point sources is significant for neutrino studies since high-energy gamma-ray and neutrino fluxes should be comparable. That the neutrino should be considered at all in the astronomical context is remarkable in view of the extreme difficulty in detecting it even in laboratory experiments. The very great amount of effort that is being expended to detect extra-terrestrial neutrinos is a measure of the importance of this new channel of astrophysical information.

The weakness of the neutrino interaction is a major disadvantage in its detection to the nuclear physicist, but it is this weakness that makes the neutrino particularly attractive to the astrophysicist. Because its presence is so difficult to detect, it is possible that a significant fraction of the energy of the universe is in this form; this aspect is of great significance for cosmology. Neutrinos are also expected to play a dominant role in nuclear reactions at the centre of stars. Because of the high densities involved, the x-rays and gamma-rays, as well as the particles, are absorbed long before they reach the surface of the star. Thus although the temperature at the centre of the sun is $\sim 10^7$ °K, the visible light that is observed is the black-body radiation from the 5000 °K surface. The mean free path of MeV neutrinos far exceeds typical stellar radii. Neutrinos are the only form of radiation that escapes from the stellar core; they provide the sole means of direct investigation of the nuclear mechanisms that are responsible for stellar energy.

The fact that energy in the form of neutrinos rapidly disappears from the region of emission makes it important as a means of rapid dissipation. As early as 1941 Gamow and Schoenberg postulated that the rapid conversion of energy to neutrinos could induce the collapse that is characteristic of supernovae explosions. As the temperature of the central region increases, the rate of neutrino emission increases, leading to a catastrophic gravitational collapse.

10.2 Neutrino physics

(A) DISCOVERY

The neutrino was first proposed as a theoretical concept in 1930 by Wolfgang Pauli. The observed features of nuclear β-decay indicated an apparent contradiction of the laws of conservation of energy and momentum; these laws have such a universal basis that, rather than accept their violation, it seemed preferable to accept the existence of a new particle, the neutrino. The boldness of this postulate is apparent when it is remembered that at this stage the only elementary particles known were the proton and electron. In 1933 Fermi developed a detailed theory of β-decay based on the existence of the neutrino, assuming the basic reactions were of the form

$$p \rightarrow n + e^+ + \nu$$
$$n \rightarrow p + e^- + \bar{\nu}$$

Although the neutrino was postulated to have negligible mass and no charge, it differed from the photon in its spin. In β-decay the energy emitted is shared between the neutrino and the β-ray, so that a spectrum of β-ray energies is expected and the conservation laws are not violated.

The Fermi theory gives a quantitative description of the interaction of the neutrino with matter; this turned out to be so weak that its detection appeared impossible. In fact it was not until twenty-three years later that Cowan and Reines succeeded in proving its existence. The reaction used was the absorption of anti-neutrinos by protons:

$$\bar{\nu} + p \rightarrow n + e^+.$$

Both the free neutron and positron have only a finite lifetime; the neutron is captured by cadmium in the scintillator giving off a detectable gamma-ray, while the positron annihilates with an electron to give two characteristic gamma-rays. These three products are easily detectable with gamma-ray counters; the appearance of three events with a predictable time sequence in shielded counters establishes the neutrino event.

The cross-section for this interaction is only 10^{-44} cm² so that very large counters had to be used; the experiment was performed beside one of the Los Alamos nuclear reactors where an intense neutrino flux $\sim 10^{13}$ MeV anti-neutrinos/cm²/sec was obtained.

(B) MUON NEUTRINO

A major advance in neutrino physics was made in 1961. For some time it had been known that the π meson and muon decayed with the emission of neutrinos: these were manifestations of the same weak interaction as the nuclear β-decay. The group of Schwartz at the Brookhaven proton accelerator succeeded in detecting the neutrino decay products of mesons produced in nuclear reactions: these neutrinos interacted with nucleons to produce muons. This interaction demonstrated that the neutrino that results from β-decay is essentially different from the meson decay neutrino. The high degree of collimation of an accelerator beam, as well as the high energy of the neutrinos, made the detection of these neutrinos somewhat simpler than in the Cowan–Reines experiment. The average neutrino flux through the 10-ton spark chamber detector was $1 \cdot 5 \times 10^3$/cm² for every 10^{11} protons striking the beryllium target. In six months of operation, 51 events were detected. This small number was sufficient to establish the existence of the muon neutrino and by symmetry its anti-neutrino; thus four neutrinos ν_e, $\bar{\nu}_e$, ν_μ and $\bar{\nu}_\mu$ are known to exist. All are of astrophysical importance, and although their origins are somewhat different, their detection presents similar problems, involving the use of very large detectors for long periods of time to detect weakly interacting particles.

(C) THE WEAK INTERACTION

The weak interaction that characterizes the production of neutrinos is associated with a coupling constant g, which gives the size of the matrix element for the following reactions:

β-decay: $\qquad\qquad n \longrightarrow p + \bar{e} + \bar{\nu}_e$

K capture: $\qquad\qquad e^- + p \longrightarrow n + \nu_e$

ν charge exchange: $\quad \bar{\nu}_e + p \longrightarrow e^+ + n$

Its magnitude is given by: $g = 3 \cdot 08 \times 10^{-12}\ \hbar^3/\text{m}^2\ \text{c}$

$$= 1 \cdot 4 \times 10^{-49}\ \text{ergs–cm}^3.$$

The cross-section for typical neutrino interactions is given by:

$$\sigma = g^2[E_\nu(c.m)]/h^4c^4$$
$$\sim 10^{-44}E_\nu(c.m)/mc^2 \text{ cm}^2$$

where the centre-of-mass neutrino energy $E_\nu(c.m) < 1$ GeV and the lepton mass is negligible. The cross-section is thus typically 10^{-44} cm^2 but increases with energy. These cross-sections are about 10^{15-20} times smaller than those for typical nuclear and electromagnetic interactions. For $E_\nu < mc^2$, the cross-section is so small that there is little hope of detection.

The weakness of the neutrino interaction makes it particularly difficult for the nuclear physicist, working with particle accelerators, to investigate. With the particle accelerators at present in existence the intensity and energy of the neutrino flux is insufficient for many experiments; with present techniques the Brookhaven and CERN accelerators produce only about twenty events per ton of detector per day. The effective upper energy cut-off is about 5 GeV. By the use of more intense beams and very large hydrogen bubble-chambers the rate of events is expected to be increased by a factor of a hundred in the next few years. The extension of the upper energy limit must await the construction of larger accelerators.

One major uncertainty in neutrino physics is the possible existence of an intermediate boson, W; this heavy particle has been postulated to explain certain features of the theory of weak interactions. It is produced and decays as follows:

$$\nu + e^+ \longrightarrow W^+ \quad \begin{array}{l} \longrightarrow \text{mesons} \\ \longrightarrow \nu_\mu + \mu^+ \\ \longrightarrow \nu_e + e^+ \end{array}$$

Despite intensive efforts, there is still no evidence for the existence of W^+. From accelerator-neutrino experiments a lower limit for its rest mass of 2 GeV is deduced. There is some hope that it may be detected in high-energy cosmic neutrino experiments.

10.3 Solar neutrinos: production

At the moment the sun is the only astronomical object which can confidently be predicted to emit detectable neutrino fluxes. These neutrinos originate in the very hot stellar core, in a volume less than a millionth of the total solar volume. This core region is so well shielded by the surrounding layers that neutrinos present the only way of directly observing it. Thermonuclear reactions are sufficiently well understood that detailed models of the solar structure can be drawn up; the values of density, temperature and composition from these estimates are believed to be very close to the actual values. A single measurement of the solar neutrino flux would enable these models to be verified. On the other hand a definite disagreement would call for a major revision of current thinking on stellar structure and evolution.

The thermonuclear reactions that are primarily responsible for the solar energy generation were proposed by Bethe in 1939. The hydrogen-burning cycle is now well-established from laboratory studies; the cross-sections, nuclear energy levels and neutrino energies are known for all the reactions. These reactions are shown in Table 10.1. Sequences II and III are alternative: the branching ratio of following II: III is 1500:1.

TABLE 10.1

Thermonuclear reactions

I		$p + p \rightarrow d + e^+ + \nu_e$	(a)
		$p + d \rightarrow \mathrm{He}^3 + \gamma$	
		$\mathrm{He}^3 + \mathrm{He}^3 \rightarrow \mathrm{He}^4 + p + p$	
	or	$\mathrm{He}^3 + \mathrm{He}^4 \rightarrow \mathrm{Be}^7 + \gamma$	
II		$\mathrm{Be}^7 + e^- \rightarrow \mathrm{Li}^7 + \nu_e$	(b)
		$\mathrm{Li}^7 + p \rightarrow \mathrm{Be}^8 + \gamma$	
		$\mathrm{Be}^8 \rightarrow \mathrm{He}^4 + \mathrm{He}^4$	
or			
III		$\mathrm{Be}^7 + p \rightarrow \mathrm{B}^8 + \gamma$	
		$\mathrm{B}^8 \rightarrow \mathrm{Be}^8 + e^+ + \nu_e$	(c)
		$\mathrm{Be}^8 \rightarrow \mathrm{He}^4 + \mathrm{He}^4$	

In each case the end result is the same $4p \rightarrow \mathrm{He}^4$. The fraction of the reaction energy, f, given off in each cycle as neutrinos, is given in Table 10.2. Also shown is the energy spectrum of the neutrinos: these are shown diagrammatically in Fig. 10.1.

TABLE 10.2

Neutrino energies

SEQUENCE	f	NEUTRINO ENERGY	MeV
I (a)	0·02	Continuous	0 → 0·4
		Line	1·44
II (b)	0·04	Line	0·38
		Line	0·86
III (c)	0·28	Continuous	0 → 14

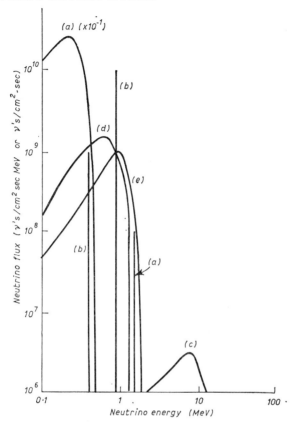

Fig. 10.1. Principal solar neutrino energy fluxes at the earth

The total energy released in the hydrogen-helium transformation is 26 MeV; 0·8 MeV is emitted on average in the form of neutrinos.

The C.N.O. cycle must also be taken into account. These reactions are illustrated in Table 10.3. The fraction f in this case is 0·06 and the neutrinos emitted from the N^{13} and O^{15} decays have a continuous

TABLE 10.3

C.N.O. cycle

$$C^{12} + p \to N^{13} + \gamma$$
$$N^{13} \to C^{13} + e^+ + \nu_e \quad (d)$$
$$C^{13} + p \to N^{14} + \gamma$$
$$N^{14} + p \to O^{15} + \gamma$$
$$O^{15} \to N^{15} + e^+ + \nu_e \quad (e)$$
$$N^{15} + p \to C^{12} + He^4$$

distribution with average energies of 0·71 and 1·00 MeV respectively. Lines are also observed but these are relatively unimportant.

Since two neutrinos are emitted whatever way the reaction $4p \to \mathrm{He}^4$ proceeds, the total neutrino flux at the earth can be determined.

Solar neutrino flux at earth, $F = 2h/z$

where $h =$ solar constant, and $z =$ average energy released in the change $4\mathrm{H}^1 \to \mathrm{He}^4$.

For $h = 2$ cal/cm²/min and $z = 26$ MeV,

$$F = 6 \cdot 5 \times 10^{10} \; \nu\text{'s/cm}^2\text{/sec.}$$

This very large flux can be broken down into contributions from the various neutrino-producing reactions. These fluxes are shown in Fig. 10.1.

Earlier models of the sun, in which the C.N.O. cycle played a larger part, were more optimistic for experimental neutrino physics since the neutrino energies are somewhat higher than in the p-p cycle and would be therefore easier to detect.

10.4 Solar neutrinos: detection

(A) RADIOCHEMICAL DETECTORS

All of the neutrinos produced in these reactions are relatively low-energy (<10 MeV) electron-neutrinos. Unlike the electron anti-neutrino which produces the easily recognizable neutron and positron, the neutrino nuclear reaction is basically

$$\nu + n \to p + e^-.$$

Neither of these products is immediately obvious in a nuclear detector against the usual sources of background: more subtle methods are therefore required.

The most promising technique for the detection of the solar neutrino flux is a radiochemical method based on the K-capture reaction $\mathrm{A}^{37} + e^- \to \mathrm{Cl}^{37} + \nu_e$ which has a measured half-life of 35 days. Pontecorvo proposed in 1946 that the inverse reaction

$$\mathrm{Cl}^{37} + \nu_e \to \mathrm{A}^{37} + e^-$$

is a feasible method of detecting neutrinos since the product, argon, can be chemically separated and the isotope can be detected by its radioactivity. The cross-section for this interaction is large but is a strong function of the neutrino energy, due largely to the possibility of transitions to excited states of A^{37}. The threshold for the reaction is 0·814 MeV so that only comparatively high energy neutrinos can be detected. The cross-section for the boron neutrinos is $1 \cdot 4 \times 10^{-42}$ cm², but that for the other neutrinos is at least 10^3 times smaller. The fraction of Cl^{37} transitions is shown in Table 10.4; although only 1 in 1500 helium nuclei are formed with the emission of a neutrino from

TABLE 10.4

Relative detection rate from various reactions

NEUTRINO ORIGIN	FRACTION OF Cl^{37} INDUCED BY THIS NEUTRINO FLUX FROM THE SUN
(a) continuous	0
line	5×10^{-3}
(b) line (0·38)	0
line (0·86)	0·075
(c) continuous	0·90
(d) continuous	4×10^{-3}
(e) continuous	$1·8 \times 10^{-2}$

boron, these high-energy neutrinos constitute 90% of the neutrinos detected by this method. The expected rate therefore is

$$R(Cl^{37} \rightarrow A^{37}) = 4 \pm 2 \times 10^{-35}/\text{target atom/sec}$$

Davis has constructed a major experiment based on this technique. The experimental method is straightforward but due to the small number of reactions expected, great care must be exercised. The chlorine used is in the form of liquid C_2Cl_4, which is a convenient form because it is readily available commercially (cleaning fluid), and has a high degree of purity. In pilot experiments 10^3 gallons of C_2Cl_4 were used; a small amount of A^{36} was added as a carrier. The detector was exposed until equilibrium was established, then helium was passed through the liquid bringing with it the A^{36} and A^{37}. The argon was separated from the helium by passing through activated charcoal. The amount of A^{37} present was estimated from its radioactivity, the Auger electrons and x-rays from the excited Cl^{37} being detected by small low-background proportional counters. The efficiency of detection of an A^{37} isotope is estimated at 44%. The major sources of background are the decay products of the interaction of the cosmic radiation with the atmosphere. By operating the experiment underground this background can be greatly reduced. In a preliminary experiment the apparatus was operated in a mine-shaft with shielding equivalent to 1800 metres of water. The rate of neutrino detection events was 0·5/day, so that

$$R(Cl^{37} \rightarrow A^{37}) < 3 \times 10^{-34}/\text{target atom/sec.}$$

This experimental limit is a factor of ten greater than the predicted flux from the sun; the majority of events detected were therefore probably due to background. This experiment had three important results:

M

(i) It showed that the background rate was not so great as to make the technique useless for astrophysics. By going to a deeper location this background could be substantially reduced.

(ii) Even the upper limit obtained enabled certain alternative nuclear reactions in the sun's core to be eliminated, e.g. less than 0·2% of the *p-p* cycles can go through the reaction

$$Li^4 \longrightarrow He^4 + e^+ + \nu(18\cdot8 \text{ MeV})$$

(iii) From this measurement an upper limit of 2×10^7 °K could be placed on the sun's central temperature. This must be compared with the predicted value of $1\cdot6 \times 10^7$ °K.

A similar experiment performed under the nuclear reactor at Savannah River, Georgia, showed that the detection technique is comparatively insensitive to anti-neutrinos, so that any ambiguity in the solar neutrino flux measurements is eliminated.

On the basis of these early experiments Davis has constructed a more elaborate detector using basically the same technique as before. The amount of detector fluid has been increased by a factor of 10^2 to 10^5 gallons; the location is now the Homestake Mine in South Dakota, where the depth is equivalent to 4700 m of water. The main detector is a cylinder of diameter 20 ft and length 48 ft. The rates expected are shown in Table 10.5.

TABLE 10.5

Expected rates in Davis experiment

Solar neutrinos	$= 7 \pm 4$ counts/day
BACKGROUND	
Cosmic-ray-produced neutrinos	$< 0\cdot2$ counts/day
Neutrinos from radioactive decay in rock	$< 0\cdot6$ counts/day
Internal α-emitters	$0\cdot02$ counts/day

The chief source of background, the neutron flux from the surrounding rock, can be reduced to a negligible amount by sealing off the underground laboratory and filling it with water.

Fowler has pointed out that this radiochemical method, which is particularly sensitive to the $B_e^7(p,\gamma)B^8$ reaction is a very accurate method of determining the sun's central temperature. The cross-section for the reaction is a very strong function of temperature

$$\sigma(B_e^7) \propto T_c^{14}.$$

An uncertainty in the Davis measurement of 50% would still give T_c to within 10%.

This detector has no directional properties so that the solar origin of the detected neutrinos must be established by studying the variation of counting rate with time. The ellipticity of the earth's orbit around the sun will give a 7% modulation of the neutrino detection rate over a year, if the flux is solar.

(B) NEUTRINO–ELECTRON DETECTORS

The most serious limitation of radiochemical methods of detection, such as that used by Davis, is that no information is obtained about the direction of the neutrino. The only energy information is that the neutrino is above threshold. Other methods, based on less certain theoretical grounds, have been developed which will provide this information. The most important of these is that proposed by Reines and Kropp in 1964. This depends on the scattering of neutrinos by electrons

$$\nu_e + e \rightarrow \nu_e + e.$$

For MeV neutrinos, the cross-section for this scattering is greater than 10^{-44} cm^2. The recoil electrons have an energy and angular distribution close to that of the incident neutrinos.

This reaction is predicted by most weak interaction theories, but it has not yet been observed in the laboratory. Although both the solar neutrino flux and the detection mechanism are unverified postulates, a detector has been built consisting of 1000 gallons of liquid scintillator viewed by 70 photomultiplier tubes. This detector is located in a deep mine in South Africa. It is surrounded by 40 tons of paraffin wax blocks as a shield. Initially only the energy of the neutrinos will be detected. The predicted rate is 20 events/year with an energy greater than 6 MeV. Hence, if successful, this experiment will provide a crude energy spectrum of the solar neutrino flux. A preliminary experiment with 200 gallons of scintillator gave an upper limit in agreement with that obtained in the Davis experiment.

Another technique, which is being pursued at the Case Institute, was proposed by Reines and Woods in 1965. Lithium absorbs neutrinos with the emission of an electron

$$\nu_e + \mathrm{Li}^7 \rightarrow \mathrm{Be}^7 + e^-.$$

The direction of the emitted electron is close to that of the incident neutrino; hence this method may be used to locate the direction of neutrino sources. Several tons of lithium are used in the form of thin slabs surrounded by 2000 gallons of liquid scintillator. The detector is located 2000 feet below the surface of the earth and is surrounded by a large particle detector which acts as an anti-coincidence shield. The rate expected is 50 events/year for energy greater than 6 MeV.

Because the threshold of the Davis experiment is lower, it is sensitive to neutrinos from either B^8 or the C.N.O. cycle. These other experiments are sensitive only to the B^8 neutrinos; hence the experiments supplement one another.

10.5 Other emission processes

Although neutrino emission from the low Z nuclear reactions in the sun provide the most optimistic prospect of a detectable neutrino flux, several other sources and mechanisms have been examined theoretically for possible large neutrino fluxes. In general the results have been negative, but it has been shown that neutrino interactions do play an important role in astrophysical processes. The most important of these is in stellar evolution where, towards the end of a star's lifespan, neutrino emission constitutes the most efficient means of dissipation of energy from the dense and very hot stellar core.

If conditions in a star's interior are ever such that large amounts of energy can be emitted in the form of neutrinos, then they constitute an energy sink and will shorten the star's lifetime. This energy sink is particularly important prior to the supernovae outburst. Before considering the fluxes expected from various astronomical sources, the principal mechanisms will be briefly considered.

(A) β-DECAY OF HEAVY ELEMENTS

In highly evolved stars significant concentrations of elements with Z greater than 16 (oxygen) are built up. Decay from certain isotopes, if present in sufficient quantities, may yield considerable neutrino radiation. These processes have been considered by several authors [Fowler (1966), Ruderman (1965)]; a convenient method of estimating the importance of these processes is to evaluate the neutrino efficiency, e, defined as

$$e = \frac{\text{energy emitted as neutrinos, } E_\nu}{\text{total rest mass energy, } Mc^2}.$$

About 2% of the energy released in the p–p chain is in the form of neutrinos; for the C.N.O. cycle this figure is 6%. Since these processes never go to completion, e for hydrogen burning is about 10^{-4}. Fowler has shown that helium burning yields a negligible amount of neutrinos, as does carbon and oxygen burning. Oxygen burning, however, takes place at T greater than 2×10^9 °K so that pair annihilation radiation is significant; therefore e is about 10^{-4}.

The production of elements with A greater than 60 should give significant neutrino emission by β-decay from neutron-rich isotopes. In practice the abundance of these isotopes is less than 10^{-6} so that e is estimated as about 3×10^{-10}. Hence even in stars more highly evolved than the sun it is unlikely that the nuclear reactions of the heavier elements can compensate for the increased distance: a star only five light-years away, by the inverse square law, has a flux 10^{-11} times that of the sun at the earth.

(B) THE URCA PROCESS

Inverse β-decay can take place if sufficiently energetic electrons are available

$$e^- + {}_ZX^A \longrightarrow {}_{Z-1}X^A + \nu_e;$$

the resultant nucleus may then undergo β-decay

$$_{Z-1}X^A \longrightarrow {}_ZX^A + e^- + \bar{\nu}_e.$$

The net result of these two processes is to convert the kinetic energy of the initial electron to a neutrino and anti-neutrino. This is an extremely efficient mechanism for losing energy provided the following conditions are satisfied:

(i) The electron energy is greater than the threshold for the inverse β-decay reaction. This may be of the order of an MeV, so that if the electron energy distribution is thermal, temperatures in excess of $10^9\,°K$ are required for a significant density of relativistic electrons.

(ii) The resultant nucleus $_{Z-1}X^A$ is unstable and decays by β-emission. At the high temperatures and densities at which condition (i) is satisfied, other processes will compete with decay; in particular photonuclear processes may deplete the $_{Z-1}X^A$ concentration.

(iii) A high density of the isotope $_ZX^A$ is present. These nuclear abundances depend on the model of nuclear synthesis assumed.

Under the conditions normally assumed to hold in the core of highly evolved sources, this process is probably not very important. According to Chiu (1963) the most significant contributors to the Urca process are the isotopes Cl^{35} and S^{32}.

(C) ELECTRON–NEUTRINO INTERACTIONS

The weak interaction theory of Feynman and Gell-Mann predicts many interactions between neutrinos and electrons, which may be important in stellar processes. Since this theory has not been fully verified, these interactions are still speculative. These processes, in each of which a neutrino pair is formed, have been reviewed in detail by several authors [Chiu (1963), Ruderman (1965)] and will be merely listed here:

(i) Neutrino bremsstrahlung, $e + X \rightarrow e + X + \nu_e + \bar{\nu}_e$. This is the exact analogue of photon bremsstrahlung with the nucleus playing a passive role. The cross-section is quite small compared with the other processes.

(ii) Photo-neutrino, $\gamma + e \rightarrow e + \nu_e + \bar{\nu}_e$. This is equivalent to the Compton effect with a neutrino pair replacing the scattered photon.

(iii) Pair annihilation, $e^- + e^+ \rightarrow \nu_e + \bar{\nu}_e$. If the temperature is sufficient for electron pairs to exist in equilibrium with thermal radiation then the electron pair is annihilated and a neutrino pair may be formed. For non-degenerate matter the neutrino emission is a strong function of temperature ($\propto T^9$).

(iv) Plasma neutrinos, Plasma $\rightarrow \nu_e + \bar{\nu}_e$. In a dense plasma, neutrino pairs are formed from plasma excitations.

The regions of density and temperature in which the above processes dominate are shown in Fig. 10.2. (iii) is most important for T greater than 10^9 °K. (iv) dominates at very high densities ($\rho \sim 10^4$ to

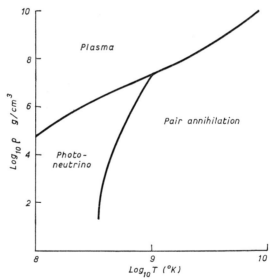

Fig. 10.2. Electron–neutrino interactions as a function of T and ρ

10^8 g/cm³) while (ii) is most important in the intermediate regions. All of these are negligible compared with visible radiation from a star with central temperature less than 2×10^7 °K; but for T greater than

5×10^8 °K, neutrino processes constitute most of the energy loss from stars.

10.6 Stellar and galactic sources

All explosive astrophysical events are potential sources of neutrinos. The uncertainty as to the mechanisms in these sources, together with the uncertainty with regard to neutrino production mechanisms, makes flux estimates highly speculative even by astrophysical standards. Three sources are principally considered: (*a*) neutron stars, (*b*) supernovae, (*c*) radio galaxies and quasars.

(A) NEUTRON STARS

This configuration, if it exists, will cool rapidly by neutrino emission. Although the initial flux will be very high ($\sim 4 \times 10^{39}$ ergs/sec for $T > 2 \times 10^9$ °K), it will last for a very short time. Since the frequency of collapse is low and the rate of collapse high, the chances of registering a collapse are small.

(B) SUPERNOVAE

Prior to the supernovae outburst, neutrino processes are believed to be important. Although the rate of energy loss by neutrino emission is probably not sufficient to destroy the mechanical equilibrium and hence trigger the collapse, it has the effect of accelerating the final evolutionary phases. As the emitted neutrino energy and the stellar densities increase, the mean free paths for ν_e and $\bar{\nu}_e$ become less than the stellar radius. At the same time the temperature becomes sufficient for ν_μ and $\bar{\nu}_\mu$ to be formed; these will escape from the star but will probably have energies too low for detection. Fowler and Hoyle (1964) have calculated that a neutrino flux of 10^{47} ergs/sec might be emitted for a few seconds prior to the actual collapse. The observed rate of supernovae outbursts in the Galaxy ($<1/30$ years) makes detection of this flux unlikely.

If the relativistic electrons whose radio emission is observed in supernovae remnants are continuously produced via meson decay from p–p collisions, then the neutrino flux can be estimated.

$$\pi^+ \longrightarrow \mu^+ + \nu_\mu$$
$$\mu^+ \longrightarrow e^+ + \bar{\nu}_\mu + \nu_e.$$

Hence three neutrinos are emitted for every charged meson produced. These fluxes should be comparable with the high-energy photon flux due to π° decay; since gamma-rays have not been detected, this mechanism of continuous electron production is in some doubt.

(c) RADIO GALAXIES AND QUASARS

The same arguments that apply to neutrinos resulting from the continuous production of relativistic electrons in supernovae remnants can be applied to extra-galactic radio sources. If the same amount of energy, as goes into high-energy electrons, goes into neutrinos of all types, then the rate of emission from a typical radio galaxy is $\sim 10^{44}$ ergs/sec. Even this large amount of energy emission over the lifetime of the radio galaxy is small compared to the total rest mass so that $e \sim 10^{-4}$.

If the process of gravitational collapse is important in quasi-stellar sources and radio galaxies, as Hoyle and Fowler suggest, then neutrino emission from the hot collapsed region will be considerable. The mechanism is the electron pair annihilation process which, because of its strong dependence on temperature, depends on the mass assumed for the object. Fowler has evaluated the following expression for E_ν:

$$E_\nu = 10^3 \left(\frac{M_\odot}{M}\right)^{3/2} c^2 \text{ ergs/g.}$$

This gives $e = 10^{-6}$ for $M = 10^6 \, M_\odot$ but at the lower mass limit (set by degeneracy), $e = 3 \times 10^{-2}$ for $M = 10^3 \, M_\odot$.

10.7 Neutrinos and cosmology

Since neutrinos, particularly those of low energies, are virtually undetectable it is natural that they should be utilized in astrophysics to bridge the gap between theory and experiment. This use of the neutrino is analogous to the situation which first led to the postulate of its existence; because β-decay appeared to violate well-established conservation laws, an 'undetectable' particle was predicted. In cosmology many theories would predict an average mass density some two orders of magnitude greater than observations indicate; this missing mass has been postulated to exist in many forms, e.g. unobservable massive objects which have collapsed to their Schwartzchild radii. One suggestion which appears to have many attractive features is that this mass is smeared through the universe in the form of a Fermi sea of low-energy neutrinos.

In most cosmological theories the value of the critical density ρ_c, defined as

$$\rho_c = 3H^2/8\pi G,$$

is fundamental. Here H is Hubble's constant and G is the gravitational constant. Using the accepted values of these constants, this gives $\rho_c \sim 10^{-29}$ g/cm³. This quantity is arrived at by considering the expansion velocity of galaxies relative to one another and the restraining gravitational influence of other galaxies; if the present universal density is greater than ρ_c, then the universe will eventually cease to expand and will contract at some stage. The alternative, $\rho < \rho_c$, implies an 'open' universe with continuous expansion.

Estimates of the contribution of various forms of mass-energy to the average density of the universe are given in Table 10.6.

TABLE 10.6

Mass density of the universe

Mass of galaxies	3×10^{-31} g/cm³
Black-body radiation	6×10^{-34} g/cm³
Inter-galactic neutral hydrogen density	10^{-34} g/cm³
Ionized hydrogen density	Unknown

Cosmology is still so speculative that too much attention should not be paid to this apparent discrepancy. Even the value of Hubble's constant or the interpretation of the red-shifts of distant galaxies is not undisputed. It is of interest to consider the consequences of the missing mass being in the form of neutrinos, i.e.,

$$\rho(\nu) = 10^{-29} \text{ g/cm}^3 \sim 10 \text{ keV/cm}^3.$$

Even the solar neutrino energy flux at the earth is 10^2 times greater than this and it is still undetected. The neutrino density is related to the Fermi energy E_f in eV by the relation:

$$\rho(\nu) = \frac{1}{8\pi^2}\left(\frac{E_f}{\hbar c}\right)^3 \frac{E_f}{c^2}.$$

For $\rho(\nu) \sim 10^{-29}$ g/cm³, $E_f \sim 0.01$ eV, too small to be significant in laboratory nuclear experiments. Hence one cannot expect very drama-

TABLE 10.7

Neutrino energy density

Distortion of the end of electron spectrum in β-decay	$\rho(\bar{\nu}_e) < 3 \times 10^{-12}$ g
Gravitational collapse of galactic scale concentration of neutrinos	$\rho(\nu) < 10^{-16}$ g/cm³
Attenuation of ultra-high-energy cosmic rays	$\rho(\nu) < 3 \times 10^{-17}$ g/cm³
Cowan–Reines experiment, $\bar{E}_\nu \sim 5$ MeV	$\rho(\nu_e) < 10^{-24}$ g/cm³
Davis experiment, $\bar{E}_\nu \sim 0.9$ MeV	$\rho(\bar{\nu}_e) < 5 \times 10^{-26}$ g/cm³

tic observational effects, particularly since these neutrinos would be almost certainly of low energies and probably below the threshold of presently conceived techniques. Ruderman (1965) has discussed the present upper limits that can be placed on the background neutrino flux. These are summarized in Table 10.7.

10.8 Muon neutrinos

Most of the discussion thus far has been in terms of the electron neutrino and anti-neutrino. The muon neutrinos and anti-neutrinos are even more an unknown quantity; because of the higher rest energy of the muon (106 MeV), they are probably only important at high energies. Again the problematic existence of the intermediate boson complicates both flux estimates and detection efficiencies. Surprisingly it is this branch of 'natural' neutrino studies that has produced the first positive results. Since the fluxes detected are almost certainly produced in cosmic-ray collisions in the atmosphere, they do not provide information of direct astrophysical significance. None the less, these difficult experiments are an important contribution to neutrino and cosmic-ray physics; in the future the elaboration of these techniques may provide surprising results.

The fluxes of neutrinos produced in the atmosphere can be estimated quite accurately from an analysis of cosmic-ray experiments, particularly from studies of the muons. The neutrinos can result from the decay of K, π or muons, so the production spectra of these particles are required. Independent estimates by various authors show that the predicted fluxes have an uncertainty of no greater than a factor of two [Acher et al. (1965)].

Two groups using essentially similar techniques have detected these fluxes. These ν_μ techniques differ from ν_e detection techniques in the following characteristics:

(i) Because of the rest mass differences of the electron and the μ-meson, the ν_μ energy range investigated is higher ($>10^9$ eV).

(ii) The target used is the earth's rock crust. Unlike the ν_e experiments, where the target is a chemically pure substance of known dimensions, there is considerable ambiguity about the nature and position of the interacting target atoms.

(iii) The characteristic of ν_μ interactions, the emission of a high-energy μ-meson, is more easily recognized than its electron counterpart; because of its penetrating power, the reaction can occur a considerable distance from the detectors.

(iv) The actual interaction whereby the μ-meson is emitted is not isolated by the detection technique; the calculated cross-sections are $\sim 10^{-37}$ to $10^{-38}\,E_\nu$, where E_ν is in GeV, for a number of interactions.

(a) Elastic collisions

$$\nu_\mu + \mathcal{N} \longrightarrow \mu + \mathcal{N} \qquad (\mathcal{N} = \text{nucleon})$$

(b) Inelastic collisions

$$\nu_\mu + \mathcal{N} \longrightarrow \mu + \mathcal{N} + \pi \text{ mesons}$$

(c) Interaction in which the intermediate boson is produced.

(v) Because the background for the experiment is cosmic-ray-produced μ-mesons, which will be attenuated with depth, these experiments must be performed at the greatest possible depths. In addition, to remove any ambiguity about the events detected, the detectors should possess some directional properties. The cosmic-ray-produced μ-mesons should be peaked towards the zenith, whereas the horizontal direction, presenting a greater target depth, should give more neutrino-produced μ-mesons. By choosing only those events close to the horizontal, the origin of the particles can be established. This directionality has the added advantage that when the experiments have run over a long enough period, anisotropies in sidereal time may become apparent. If such an anisotropy were to appear it would indicate an extra-terrestrial ν_μ source superimposed on the cosmic-ray terrestrial neutrino fluxes.

The Kolar Gold Fields neutrino experiment in India is in one of the deepest mines in the world, at a level equivalent to 7·5 km of water cover (7600 ft). Three telescopes were initially used, each consisting of two vertical slabs of scintillator (2 m $\times$ 3 m) separated by 80 cm. Between the scintillator detectors there were stacks of neon flash tubes which enabled the meson trajectory to be estimated to within 2° in zenith angle; the resolution in azimuth was poor.

Preliminary results published in 1965 [Acher *et al.* (1965)] showed a total of 13 events over an effective running period of 234 days. Eight of these events were rejected (from the arrival direction) as due to atmospheric muons. The ν_μ rate deduced was $1·2 \times 10^{-12}$ particles/cm² sec sterad. This rate was somewhat higher than expected for elastic and inelastic collisions which might indicate the existence of the intermediate boson or of an extra-terrestrial ν_μ contribution. The former view is supported by one event in which two particles traversed the detector. This might correspond to an event of the type

$$\nu_\mu + p \longrightarrow p + \mu + W$$
$$W \longrightarrow \mu + \nu_\mu.$$

Subsequent results obtained with an enlarged and improved detector have not confirmed this result.

The collaborative experiment of the Case Institute, U.S.A., and the University of Witwatersrand, South Africa [Reines et al. (1963)], is located in the 10,500 ft. level of a South African gold mine. Using 36 large liquid scintillation detectors ($5 \cdot 5$ m $\times$ 58 cm $\times$ $12 \cdot 5$ cm) the effective area is three times greater than the Kolar Gold Field experiment, but the angular resolution is poor. The effective target is estimated as 10^3 tons of rock. Ninety-four days of operation have yielded 7 neutrino-events, giving a rate of $0 \cdot 6 \times 10^{-12}$ particles/cm^2 sec sterad in agreement with the above.

A third large experiment, which is now in operation at the University of Utah [Keuffel et al. (1965)], is composed of a complex of spark and Cherenkov counters immersed in 2000 tons of concrete. Only mesons travelling upwards will be registered, but their trajectories will be located to within a fraction of a degree.

One other experiment which is producing interesting results is worthy of mention here although it is not certain that the events recorded are due to neutrinos. Cowan and his collaborators (1965) at the Catholic University of America, Washington, have for some years been detecting μ-mesons produced in a massive counter by neutral primaries. Their detector and its location have been changed several times. The present version consists of a spark chamber of side 4 ft which is surrounded by an anti-coincidence counter shield. μ-mesons of energy $\sim$100 MeV are identified by their decay pattern and the tracks in the spark chamber are used to draw a map of the points of origin of the primaries on the celestial sphere. Several strong peaks occur when these points of origin are correlated with sidereal time, although they do not appear on a solar plot. Some of these peaks are greater than four standard deviations above the general background recorded. The peaks have not been correlated with any astronomical objects nor has the nature of the particles, which penetrate the anti-coincidence shield without triggering it and produce a μ-meson, been established. If they are neutrinos, then the fluxes are surprisingly large, and detectable effects would have been expected in the Kolar experiment. If not neutrinos, then a cross-section of 10^{-28} cm^2 is indicated.

10.9 Astrophysics and neutrinos

It is of interest that astrophysical studies have provided some important pointers to neutrino physics. The proof of the existence of the intermediate boson may come from these studies; if its rest mass exceeds the energies available in the next generation of accelerators, then the natural neutrino flux may be the only possible source of these interesting particles for many years. The values of charge, magnetic moment and charge radius of the neutrino are still unknown; Bernstein, Feinberg and Ruderman (1963) list the upper limits to these values deduced from astrophysical arguments. These values together with the upper limits from other sources are given in Table 10.8.

TABLE 10.8

Neutrino properties from astrophysics

PROPERTIES	ν_e	ν_μ
Charge (electronic charge)	$<10^{-13}$ from astrophysics $<4 \times 10^{-17}$ from charge conservation	$<10^{-13}$ from astrophysics $<3 \times 10^{-5}$ from charge conservation
Magnetic moment (Bohr electron magnetons)	$<10^{-10}$ from astrophysics $<1 \cdot 4 \times 10^{-9}$ from absence of electron scattering	$<10^{-10}$ from astrophysics if $m(\nu_\mu) < 1$ keV $<10^{-8}$ from absence of pion production
Charge radius cm)	$<4 \times 10^{-15}$ from astrophysics $<4 \times 10^{-15}$ from absence of $e - \nu$ scattering	$<4 \times 10^{-14}$ from astrophysics if $m(\nu_\mu) < 1$ keV $<10^{-15}$ from absence of pion production

Chiu (1963) has made a detailed analysis of the way in which neutrino processes affect the rate of evolution of stars. Taking as a starting point the observed gap in the H.R. diagram between Main Sequence stars and white dwarfs, he concludes that the accelerated evolution in this region is proof of the existence of the direct electron–neutrino interaction. From this a lower limit of the coupling constant of the interaction can be set at only one half the theoretical value.

11
Microwave Background

11.1 Cosmology

Since cosmological theories can seldom be confronted with observation, it may almost be said that there are as many cosmological theories as there are cosmologists. For the outsider there is very little to choose among the various theories since all have some satisfactory aspects; the few observational facts that are available can be explained by a large number of theories which are contradictory in other respects. The selection of one such theory as a basis for work in some branch of astrophysics inevitably reflects some subjective bias. There is no simple cosmological theory, in the sense that no theory has yet been devised that does not require the inclusion of some concepts alien to our conventional laboratory physical concepts. Whether cosmology requires that the laws of physics, as deduced in the laboratory, must be extended or modified, or whether the physical laws, as we know them, are the basis of all phenomena on a nuclear, as well as cosmological, scale, is the principal reason that this speculative field has attracted the attention of the most prominent scientific minds of our time. Certainly no problem can be so great as that of understanding the evolution of the universe; for all that the problem is still very far from solution.

Hoyle (1965) has pointed out that three alternative possibilities arise at an early stage in any cosmological theory:

(i) That the laws of physics do not hold true for all time: at an earlier epoch different laws may have been in force.

(ii) That the universe was created a finite time ago.

(iii) That the distribution of matter in the universe, as observed now, has not always been the same.

The third alternative has been the one most seriously considered. Cosmological theories can be classified under three headings:

(1) *The Evolutionary theory*. For many years this was the orthodox theory. The observed expansion arises from a single large explosion. The expansion will continue indefinitely with the continuous dilution of the energy density.

(2) *The Steady-State theory*. Matter is continuously created so that the average energy density remains constant. As galaxies expand, matter is created to fill the gaps. Matter eventually condenses to form galaxies, so that the continuing birth of galaxies should be observed. In what form matter is created is a fundamental difficulty since all the forms suggested should have some observable properties. Aesthetically pleasing, this theory implies the breakdown of one of the most fundamental conservation laws of physics.

(3) *The Oscillating Universe*. This theory is essentially a modification of the Evolutionary theory, in which it is postulated that the observed expansion will eventually cease and a contraction take place. When matter has condensed to a single massive object, then another explosion will occur and the cycle will begin all over again. Currently the Steady-State school of thought is in disfavour; there is as yet no observational evidence to favour (3) rather than (1).

11.2 Black-body radiation prediction

For the oscillating universe theory to be a complete cyclic process it is necessary to suppose that as a result of the collapse matter will be transformed to the state it was in when the expansion stage of the universe began. Hence the nuclear synthesis that has been going on in the universe must be reversed and all matter decomposed again into its constituent nucleons; the temperature at which this would occur is $\sim 10^{10}$ °K. Dicke and his collaborators (1965) have considered the consequences of this high temperature; they point out that this treatment is not confined to the Oscillating Universe model but could also occur in the initial 'big bang' explosion.

At a temperature of 10^{10} °K, the characteristic thermal photons have energies $\sim mc^2$ so that electron–positron pair production and annihilation processes would be very important and would be in thermal equilibrium: the electron neutrino–anti-neutrino production processes would lead to a neutrino density also in thermal equilibrium.

The important feature of this high temperature is that the thermal black-body spectrum would persist long after the universe had begun to expand again, although expansion would cause a shift of the black-body temperature. The black-body spectrum would thus be shifted through many orders of magnitude from the gamma-ray region of the electromagnetic spectrum to much lower frequencies, probably in the radio region. Since the exact conditions in the initial fire-ball cannot be reliably estimated, it is difficult to estimate the present black-body temperature. An upper limit can be obtained by supposing that the maximum energy density deduced for the universe, 2×10^{-29} ergs/cm³, be in the form of thermal radiation; this would correspond to a maximum black-body temperature of 40 °K. Measurement at a frequency of 404 Mc/s give an upper limit <16 °K.

In the belief that this remnant radiation, if it existed, might be observable, Dicke *et al.* devised an experiment to search for it in the microwave region of the spectrum; while their experiment was in course of construction, the first indications of the existence of this black-body background were reported.

It is of interest that some twenty years earlier Gamow had suggested the existence of this black-body radiation remnant of a big bang. Dicke *et al.* were unaware of this suggestion.

11.3 Observation

In an experiment whose importance was scarcely recognized at the time, Penzias and Wilson (1965) made the first direct observations of the remnant black-body radiation. Using a 20-ft horn-reflector antenna at the Bell Laboratories in the United States, they studied the noise temperature at the zenith. Their detector was a travelling-wave maser; a liquid helium-cooled reference termination was alternated with the antenna as a calibration. The frequency investigated was 4080 Mc/s ($\lambda = 7.4$ cm).

In an experiment of this sort, where an isotropic background is suspected, it is most important that the results be corrected for all instrumental effects. This is most difficult here since there are no known celestial point sources against which a calibration may be made. If the background is of solar or Galactic origin, then a modulation of the signal would be expected in solar or sidereal time; an extra-galactic background would be isotropic in most cosmologies.

The measured antenna temperature was 6·7 °K; the error in this measurement was estimated at ±0·3°, arising from uncertainty in the reference termination. The principal contribution to this antenna temperature was from atmospheric absorption; this is a well-known effect. Its importance was measured by varying the zenith angle of the antenna and recording the change of temperature with angle. Other losses arose from the antenna itself: these were either measurable or could be computed.

The contributions to the antenna temperature are summarized in Table 11.1 together with the errors in their estimation:

TABLE 11.1

Experiment temperatures

Total antenna temperature	$= 6\cdot 7^\circ \pm 0\cdot 3^\circ$
Atmospheric absorption	$= 2\cdot 3^\circ \pm 0\cdot 3^\circ$
Ohmic loss in antenna	$= 0\cdot 8^\circ \pm 0\cdot 4^\circ$
Back-lobe response of antenna	$= {<}0\cdot 1^\circ$
Sum	$= 3\cdot 2^\circ$
Discrepancy	$= 3\cdot 5^\circ$

The error in the discrepancy of $3\cdot 5^\circ$ was put at ± 1 °K. Penzias and Wilson concluded that they had detected a source of radiation which, within the limits of their experimental accuracy, was isotropic, unpolarized and free from seasonal variations.

11.4 Interpretation

The results of Penzias and Wilson were published simultaneously with the prediction of Dicke *et al.* They tentatively interpreted the results as the predicted black-body spectrum. This interpretation has some important consequences for cosmology. Peebles (1967) pointed out that this temperature, plus the observed abundance of helium, leads to an upper limit of the present matter density of the universe.

The scheme for the thermal history of the universe as outlined by Dicke *et al.* is shown in Fig. 11.1. At a temperature of 10^{10} °K, matter and thermal densities are about equal and are $\sim 10^{6-7}$ g/cm^3. Rapid evolution takes place with radiation densities predominating. After about 10^{3-4} years, matter density takes over until we reach the present situation where there is a large difference between the two. At an early stage deuterium would be formed; this stage would not be expected until the temperature was low enough for the deuterium to be stable against photo-disintegration. If the density is sufficiently high at this point then significant amounts of helium will be formed. This reaction will cease when the temperature falls. Hence from the observation of the present helium content in the universe, an upper limit to the matter density at the time of helium formation can be deduced. This, taken with the present black-body temperature, permits the present matter density to be predicted assuming a cosmological model.

There is considerable controversy about the present helium content of the universe. If a value of 25% by mass is accepted, then Peebles deduces an energy density of 3×10^{-32} g/cm³, considerably smaller than estimates by other methods. This estimate is based on Einstein's general relativity; it apparently implies an open universe, the matter density deduced being 600 times too low for that required for a closed universe.

For an oscillating, and hence closed, universe, it is necessary to

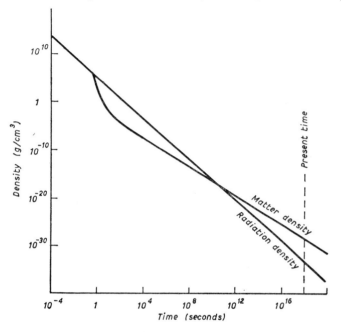

Fig. 11.1. Possible history of the universe

assume a modified form of Einstein's field equations. In the Brams–Dicke cosmology, the evolution of the universe would take place sufficiently rapidly for negligible amounts of helium to be formed. Hence a matter density of 2×10^{-29} g/cm³ at present would be compatible with both the 3·5 °K field and the observed helium content. This still leaves the problem of what form this matter density can take, since it is unobservable with present techniques.

11.5 Subsequent experiments

The single observation of Penzias and Wilson of this surprisingly large microwave flux was not sufficient to confirm the thermal nature of the spectrum. The large flux indicated was clearly above the extrapolated Galactic background, so that whatever processes were involved, a spectrum that rose rapidly with wavelength was indicated. The exact shape of the spectrum could only come from a number of observations at different wavelengths; because of the importance of the Penzias and Wilson result, these observations were rapidly forthcoming.

The first confirmation came fittingly from the group who had first made the prediction. Roll and Wilkinson used an experimental arrangement similar in principle to that of Penzias and Wilson, but at a shorter wavelength ($\lambda = 3\cdot2$ cm). Their results indicated a black-body spectrum with temperature $3\cdot0 \pm 0\cdot5$ °K. Isotropy was established to within 10%. Subsequently Penzias and Wilson revised their result at $7\cdot4$ cm to $3\cdot3 \pm 1$ °K.

An ingenious, but indirect, method of measuring the flux at shorter wavelengths was then proposed. Some twenty-five years ago peculiar absorption lines were found in the spectra of the stars ζ Ophiuchi and ζ Persei. These lines could be explained as absorption by CN molecules; but since they were from an excited level, it was necessary to postulate a temperature >2 °K in the region where the absorption took place. This result remained unexplained until the Penzias and Wilson result was published. Then, by reprocessing the previous observation and making fresh observations, Field and Hitchcock (1966), and Thaddeus and Clauser (1966) estimated the temperature necessary to explain the observed absorption. At $\lambda = 2\cdot6$ mm a black-body temperature between $2\cdot7$ and $3\cdot4$ °K was deduced. This point remains the shortest wavelength measurement of the black-body spectrum. The measurements to date are summarized in Table 11.2 and Fig. 11.2.

TABLE 11.2

Black-body temperatures

	λ (cm)	T_{BB} (°K)
Penzias and Wilson	21·1	$3\cdot1 \pm 1\cdot0$
Howell and Shakeshaft	20·7	$2\cdot8 \pm 0\cdot6$
Penzias and Wilson	7·4	$3\cdot3 \pm 1\cdot0$
Roll and Wilkinson	3·2	$3\cdot0 \pm 0\cdot5$
Welch *et al.*	1·5	$2\cdot0 + 0\cdot8$ $- 0\cdot7$
Stokes *et al.*	0·86	$2\cdot56 + 0\cdot17$ $- 0\cdot22$

All the points are consistent with a black-body temperature of about 3 °K. However, no observations have been made at wavelengths below the predicted maximum of the spectrum. These measurements are particularly difficult to make because of strong atmospheric radiation. An

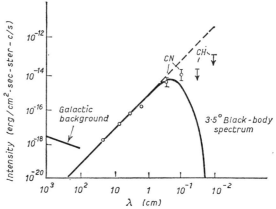

Fig. 11.2. The 3·5° black-body microwave spectrum

upper limit is however available from a study of CH lines. Since no anisotropy has been detected in the microwave flux (less than 0·1% variation in all directions) the interpretation of the observations as a 3 °K black-body radiation spectrum, which pervades the universe, is generally accepted.

Further experiments in this field will be concentrated in the following fields:

(i) Refinement of the above experiments to give a more accurate value to the black-body temperature.

(ii) Development of balloon and satellite experiments to make the difficult, but important, observations in the millimetre region.

(iii) Confirmation of the isotropic nature of the flux. A small anisotropy could have important consequences for cosmology. In particular, a small 24-hour variation is expected due to the sun's motion if the microwave flux is universal.

(iv) Refinement of methods of estimating helium content of the universe. In particular the spectra of old stars, which should have low metal abundances, but helium abundances reflecting the helium production in early stages of the expansion of the universe, will be studied.

11.6 Astrophysical consequences

Apart from the evidence for the primeval fire-ball that the 3 °K black-body field provides for cosmology, the existence of this field has other important astrophysical consequences. These consequences are independent of the origin of the microwave flux, i.e. whether it is the remnant of the primeval fire-ball or whether it has some other origin. The importance of this flux follows from consideration either of (*a*) the photon density $\sim$500/cm³, or (*b*) the energy density 0·7 eV/cm³; these values are contrasted with estimates in other parts of the spectrum in extra-galactic space in Table 10.6.

Greisen (1967) has classified the consequences under two headings: (i) absorption effects, (ii) production effects. These effects are related since absorption of energy by one process implies that energy must reappear in one of the production processes. For convenience they are considered separately.

The principal absorption processes are:

(1) Energy drain of relativistic electrons by the Compton effect.
(2) High-energy photon absorption by photon–photon pair production (9.5).
(3) Disintegration of high-energy heavy cosmic-ray nuclei.
(4) Meson production by ultra-relativistic cosmic-ray protons (4.7).

The principal production effects are:

(1) X-ray and gamma-ray photon production by the Compton effect (9.6).
(2) High-energy neutron production from the break-up of heavy cosmic-ray nuclei.
(3) Generation of an electromagnetic cascade on a galactic scale with the photon–photon pair production and Compton processes taking the place of photon–nuclear pair production and bremsstrahlung.
(4) Neutrino production by electron–photon interactions.

12
Pulsars

No previous work either experimental or theoretical had prepared the astronomical community for the discovery that the Galaxy is liberally populated with strange compact objects, which are almost undetectable optically, but which radiate strongly at radio wavelengths in bursts which are separated by characteristic periods constant to one part in a hundred million. The high regularity of the strongly polarized radio pulsation period is matched by the irregularity of the pulse amplitude. The discovery of the first pulsar was announced in February 1968 [Hewish, Bell, Pilkington, Scott and Collins (1968)]; in the following months there was hardly an observatory in the world that did not have an instrument trained on one of these objects. At the time of writing (January 1969) more than one hundred papers have been published on this subject. The topic can be expected to develop rapidly in the next few years. This chapter is included as a brief summary on the phenomenon as it appears today, since the nature of these objects is such that no current work on astrophysics would be complete without a brief account of their properties.

12.1 Discovery

As with many significant discoveries the first detection of a pulsar was somewhat fortuitous: it required the combination of a new technique, a new instrument and a conscientious group of experimenters who were prepared to track down the origin of a mysterious, but weak, source of interference. In 1964 Cambridge radio astronomers had found that compact radio sources, i.e. those with angular diameters less than a second of arc, exhibited rapid fluctuations in intensity. It was soon realized that this phenomenon was not intrinsic to the source but was exactly analogous to the scintillation of star-light in the earth's atmosphere which produces the familiar twinkling. In the case of the radio sources the scintillating medium is the solar wind, the thin plasma blown out by the sun. Since quasars are characterized by their small angular diameters relative to radio galaxies this phenomenon can be used to make a preliminary classification of radio sources. To exploit this technique the Cambridge radio astronomy group, under the direction of Hewish, constructed a very large telescope designed to work at long wavelengths (3·7 m), at which the scintillations are most marked. The telescope consisted of 2048 dipole antennae which covered an area of 4·5 acres. Although the array was fixed, the beam could be shifted in the north–south direction by phasing the antennae; the earth's rotation allowed the east–west direction to be scanned.

In July 1967 a sky survey was commenced; this was the first time the sky had been surveyed systematically with an instrument of such sensitivity at these wavelengths. Because the object of the search was to detect scintillations, short time constants were used (unlike the usual long integration times used by radio astronomers) and the same regions of the sky were repeatedly surveyed to detect changes in the scintillations from particular sources as the earth moved relative to the sun.

Early in the observing programme a peculiar source of interference was noted. This occurred during the night when the scintillations are normally weak; unlike the scintillations from a quasar this source of interference only lasted for a short period of time compared with the time required for a source to cross the beam of the telescope. Closer examination showed that the signal came from a fixed direction in stellar space and that the signal consisted of pulses of 20 msec duration with a regular spacing of a little over a second. So surprising was this result that initially the experimenters considered the possibility that the signal might originate from intelligent beings somewhere in the universe. The object was called CP1919 (Cambridge Pulsar at right ascension 19 hours 19 minutes).

The following properties were noted in the paper announcing the discovery:

(*i*) Because of the short time spread of the pulses the emitting region must be small. If its dimensions were of the order of $R_\odot$, then the minimum pulse width would be about one second. The source therefore appeared to be either a planet or a very compact star.

(*ii*) Simultaneous observations at different frequencies showed that the pulse arrival times were different for each frequency; at the highest frequencies the pulses arrived first. This effect could arise from dispersion in interstellar space; the velocity of electromagnetic waves is a function of the free electron density n_e in the medium traversed. The time delay in the pulses at two frequencies depends on the frequencies and the integrated electron density $\int_0^L n_e \, . \, \mathrm{d}l$ for a source at a distance L. If n_e is known, then the distance to the source can be calculated. Usually a uniform density of $n_e = 0 \cdot 1 \ \mathrm{cm}^{-3}$ is taken; in this case the distance to CP1919 is 130 pc. This would make the pulsar a relatively close neighbour of the sun.

(*iii*) If correction is made for the motion of the earth in its orbit which produces an annual change in the observed pulsar period by the Doppler effect, the pulsations have a period which is constant to one part in 10^8. (For CP1919 this period is $1 \cdot 337301109$ seconds.)

(*iv*) The amplitude of the signal varies from pulse to pulse in a random manner.

If CP1919 was a natural phenomenon, Hewish and his collaborators reasoned that it should not be unique. A re-examination of their data showed evidence for three other pulsars; CP0834, CP0950 and CP1133. They suggested the oscillations of white dwarfs or neutron stars as a possible explanation for these extraordinary observations.

12.2 Radio observations

(A) DISTRIBUTION

With 28 pulsars now listed, the distribution is consistent with a Galactic origin. A complete sky survey has not yet been made and the list so far has been compiled by many groups at different frequencies. Ten pulsars have been observed at the Molonglo Radio Observatory in Australia working at a frequency of 408 Mc/s. They surveyed a wide area of sky and found a definite tendency for the sources to lie close to the Galactic plane. In particular they found that the sources had a tendency to cluster together and to lie close to the tangential points of the spiral arms.

Two pulsars have been discovered which lie close to the Crab Nebula. One of these has the longest period yet detected (3·7 sec) and the other the shortest (30 msec). The latter has been definitely identified with the Crab, the former may lie 1·5° away. The pulsar PSR0833–45 lies within the extended radio source, Vela X, which is the remnant of a Type II supernova. PSR1749–28 is only 2° from the Galactic centre whilst CP0950 could be associated with the x-ray source LEO XR–1.

(B) DISTANCE

Estimates of the distance to the pulsars depend critically on the free electron density along the line of sight to the source. These electrons are the result of the ionization of the hydrogen atoms in the Galactic disc; the hydrogen density can be measured and if the temperature is known (from star-light), the electron density n_e can be estimated. For $n_e = 0 \cdot 1$ cm^{-3}, which is the usual value taken, distances based on the measured dispersion are given in Table 12.1. The time delay at a frequency ν is proportional to ν^{-2}. For CP1919 this gives an integrated electron density $\int n_e . dl = 12 \cdot 55 \pm 0 \cdot 06$ pc cm^{-3}. If the characteristic distance to the pulsars is 100 pc, then the Galactic density of these sources is about 10^{-6}(pc)$^{-3}$.

Some authors believe that the value of n_e has been overestimated and that the distribution is highly irregular with an average value of $0 \cdot 02$ cm^{-3}. Measurement of the absorption of the 21-cm line by neutral hydrogen supports this view. The absorption is measured in the continuum of the pulsar pulse; because the intervening hydrogen clouds are moving, the absorption line is Doppler-shifted. By measuring this shift the distance to the furthest hydrogen cloud can be calculated and thus a lower limit to the distance to the pulsar arrived at. In the case of CP0328 the source lies beyond clouds with velocities of -58 km sec^{-1}, i.e. a distance of $4 \cdot 2$kpc from the solar system. For this source the dispersion measurement gives $\int n_e . dl = 26 \cdot 75$ pc/cm^{-3} which combined with the above lower limit to the distance gives $n_e = 0 \cdot 006$ cm^{-3}. If this distance is accepted then the pulsar lies in the Galactic halo; if this is typical of all pulsars then the pulsar density in the Galaxy is considerably reduced.

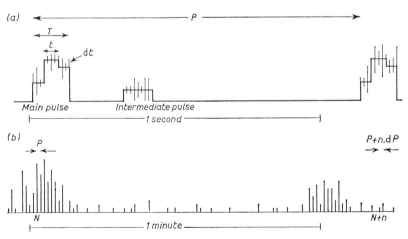

Fig. 12.1. Simplified representation of time constants associated with radio pulsations on (a) one second, and (b) one minute time scale

(c) TIME CONSTANTS

Drake (1968) has listed six time-scales that can be associated with each pulsar. These are illustrated in Fig. 12.1.

(i) *Pulsation period, P.* The most important time constant is the basic pulsation period which is characteristic of each source. Periods have been detected which range from tens of milliseconds to seconds (Table 12.1). For some pulsars the periods are known to nine significant figures. This is now called the Class I period.

TABLE 12.1

Properties of some important pulsars [Cameron and Maran (1969)]

SOURCE	P (sec)	$dP/P \times 10^{15}$	T (msec)	L (pc)
CP0328	0·714518603	—	7	270
NP0527	3·74549*	—	190	490
NP0532 (Crab)	0·03309014*	4×10^2	9·5	560
CP0808	1·292241325	—	90	57
PSR0833–45 (Vela X)	0·089208296	$1·2 \times 10^2$	2	630
CP0834	1·2737631515	$5·0 \pm 0·8$	38	128
CP0950	0·2530650372	$0·3 \pm 0·1$	21	30
CP1133	1·1879109795	$4·1 \pm 0·5$	43	49
CP1919	1·337301109	$1·1 \pm 0·5$	40	125

* Not corrected to barycentre of solar system.

(ii) *Change of period, dP/P.* The period P must be corrected to allow for the motion of the earth relative to the source which gives a progressive change resulting from the Doppler effect. When this is corrected the period was found to remain constant to one part in 10^8. However after nine months of observation on the first pulsars discovered, the Cambridge and Jodrell Bank groups detected a small but definite increase in period for all of the sources under investigation. At the same time a much larger change was observed in the pulsar associated with the Crab Nebula. Some of these changes are listed in Table 12.1 as dP/P, the change of period per period. The rate of change of the period is consistent with a linear variation with time. If the change was associated with an error in the source position (and hence an incorrect earth Doppler correction) or with a Doppler shift within a multiple component source (e.g. binary or planetary system) then the rate of change should be non-linear. The linear rate of change implies a change in the physical conditions in the source. By extrapolation of this rate of change, the lifetime of the pulsars is typically 10^7

years; for the Crab the greater rate implies a lifetime of only 2000 years.

(*iii*) *Pulse width, T.* The pulse width varies from 3 to 100 msec. It is normally about 7 per cent of the pulsation period but values from 1 to 10 per cent are observed. The pulse width has a weak inverse relation to frequency. Craft and Comella (1968) have shown that for CP1133 between 40 and 430 Mc/s

$$T \infty \ \nu^{-0.25}.$$

(*iv*) *Pulse structure.* The variations in individual pulse shapes are so diverse that no definite structure can be seen. If the pulses are averaged

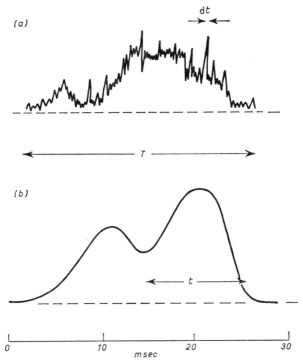

Fig. 12.2. (*a*) Typical appearance of a single pulse with good time resolution (*b*) structure of pulse averaged over many pulses

over a sufficiently long time interval, then a definite structure appears which is characteristic of each pulsar (Fig. 12.2). This structure consists of one, two or three sub-pulses, *t*, which are typically 10 msec long. This Class II period is independent of frequency. Rickett and Lyne (1968) found an intermediate pulse from CP0950 much weaker than the main pulse, occurring 100 msec before it.

(*v*) *Sub-pulses,* d*t*. As the time resolution of detection techniques has improved the pulsar pulses have revealed a complicated sub-structure. It appears that all of the pulses may be the summation of sub-pulses with time-scales less than 100 μsec.

(*vi*) *Pulse amplitude variations.* Although individual consecutive pulses vary in a random fashion, the pulse intensity does appear to wax and wane with time scales of the order of minutes to hours. Salpeter (1969) has shown that these effects could be caused by scintillations in the irregular interstellar plasma along the line of sight to the source. This plasma could either be a weak plasma in interstellar space or a denser plasma close to and associated with the source. If the plasma is far enough away from the pulsar then it will produce a measurable time delay between pulses recorded at radio observatories separated by a few thousand miles. Preliminary observations have detected no delay. The variations from pulse to pulse must occur in the pulsar itself and these minute-hour variations probably close to it.

Hewish (1968) has suggested that there may also be variations in the pulse amplitudes with time constants of the order of weeks. CP1919 appeared to show a systematic variation with a six-month period.

(D) RADIO SPECTRUM

Because of the variations of amplitude with frequency and time, the frequency spectrum of the pulsars is difficult to measure. In general it is a power law with an exponent between −1 and −2. Between 100 and 600 Mc/s the pulsar in the Crab has the form $F(\nu) \propto \nu^{-2}$.

(E) POLARIZATION

The pulses are strongly polarized. At 430 Mc/s strong elliptical polarization has been observed with variations from circular to linear. The polarization vector can change by as much as 180° from pulse to pulse. Even the very short sub-pulses exhibit rapid changes in polarization.

(F) INTENSITY

The pulse intensity varies greatly. At about 100 Mc/s, pulses of 10 f.u. have been recorded. The average pulse is about ten times less than this. Since the pulsar is radiating for less than 10 per cent of the time, the average power from the sources is well below the detection limit of conventional radio telescopes. However, once the period of the source has been established and the detection system locked in phase with it, detection is relatively easy. It is possible that some of the weaker sources located in the radio source surveys are in fact pulsating radio sources.

If the distance to CP1919 is 130 pc, then the radiated power, if isotropic, is about 10^{28-29} ergs/pulse.

12.3 Optical observations

Some of the pulsars have now been located to within five seconds of arc; this small error circle would normally be sufficiently accurate for an optical identification. A blue star of 19th magnitude lies within the error circle of CP1919 but the star is in no way unusual and the co-incidence is probably fortuitous. Apart from the Crab Nebula it appears that the pulsars are not visible on photographic plates; for CP0950 and CP1133 it is estimated that there is no optical object associated with them with a visual magnitude greater than $+21.5$.

A number of attempts have been made to detect optical pulsations from the pulsars. Many of the experiments, using a variety of techniques, concentrated on CP1919 and its blue star. At one stage a sinusoidal optical variation with a period half that of the radio pulsations was detected from this source but this result was not confirmed. It now appears that the optical pulsations from most of the pulsars are less than a few per cent of the luminosity of a 19th magnitude star.

By the end of 1968 it appeared that the pulsar phenomenon was unobservable optically. This position changed dramatically when on 15 January, 1969, three young astronomers, Cocke, Disney and Taylor of the Steward Observatory in Arizona, directed their attention to the central regions of the Crab Nebula. Using a 36-in. telescope (a modest instrument by present-day standards) they concentrated their attention on some faint stars which are barely visible in the nebulosity. In this region a small and anomalous low-frequency radio source is known to exist. Using an aperture of 22 seconds of arc and analysing for periodicity at the characteristic radio period, they saw a definite optical pulse structure. By superimposing pulses for a few minutes they could see a main pulse with a half width of about 1.4 msec followed by a smaller pulse of half width 3 msec. The separation between the pulses was 14 msec and the ratio of pulse intensities 1.7. The same pattern is observed at radio wavelengths although in that case the pulse widths are greater.

Within a few days these observations were confirmed by astronomers at the Kitt Peak and McDonald Observatories. The position of the source was definitely identified with the 'south preceding' star which was proposed by Baade in 1942 as the source of continuing activity within the Nebula. He gave its visible magnitude as $+15.45$ but although the star seemed somewhat unusual it was not generally considered as the Crab energy source. The Steward Observatory astronomers found that the visible magnitude in the main pulse during the pulse was $+15.1$; if both pulses were averaged over all time, then the magnitude was $+17.6$. It is not clear whether the star has also a non-pulsing optical component; if there is, it is small so that the star seems to have become fainter over the last 27 years. The observed optical pulses correspond to an optical flux of 10^{-13} ergs cm^{-2} sec^{-1}. The radio pulses correspond to a flux of 6×10^{-14} ergs cm^{-2} sec^{-1}. The pulsed x-ray emission is less than 5×10^{-9} ergs cm^{-2} sec^{-1}. The pulsed optical luminosity is about 3×10^{33} ergs sec^{-1}.

The spectrum of the pulses has been studied but it is difficult to distinguish any features above the continuum. This star is particularly difficult to study since its spectrum is contaminated by the spectrum of the nebula. The pulse continuum spectrum is not incompatible with the radiation being magnetic bremsstrahlung.

12.4 X-ray and gamma-ray observations

When neutron stars were first proposed it was believed that the x-ray techniques provided the best chance of experimentally detecting them. With the discovery of the pulsars there has been a revival of the neutron star concept and considerable interest in the detection of either a steady or pulsating x-ray component from these sources. There is an x-ray source in the Crab Nebula but since its dimensions are large it cannot be directly associated with the pulsar. An experiment to look for pulsations in the low-energy x-ray emission from the Crab gave an upper limit to pulsed emission as less than 5 per cent of the continuous emission. In the 20 to 50 keV region the pulsed emission upper limit from CP1919 is 1.2×10^{-4} photons sec^{-1} cm^{-2} keV^{-1}. CP0950 lies within the error circle of the x-ray source LEO XR–1 at high Galactic latitude. If the x-rays from pulsars are pulsed, then it is possible that they can be detected at ground level by analysing the optical fluorescence in the upper atmosphere for the pulsar periodicity.

Two groups using the atmospheric Cherenkov technique to search for anisotropies in the cosmic ray air shower distribution have observed tentative evidence for the emission of 10^{12-14} eV, gamma-rays from CP1133. These results have not been confirmed at lower energies.

12.5 Theories

(A) SOURCES

The fast rise time and sub-pulse structure of the radio pulsations immediately suggest that the dimensions of the radiation source must be small. If all of the emitting region is to be involved in these fast components then its dimensions must be of the order of 100–1000 km from the velocity of light limitation. These dimensions are only a fraction of the radius of a typical star. There are only three sources which have dimensions small enough: planets, white dwarfs and neutron stars. It is worth noting that because of the small size it has always been assumed that the pulsars are stellar, rather than galactic, in scale and are thus within the Galaxy.

White dwarfs are one of the last stages in stellar evolution, in which stars, which have exhausted their nuclear fuel, contract to a relatively stable, but cold, state. In this state gravitational forces are balanced by the degenerate electron pressure in the interior. The radius is about that of a planet and the central densities can be as high as 10^{10} g cm^{-3}. White dwarfs have been observed optically as faint stars so their existence is well established.

On the other hand neutron stars are still very much at the conjectural stage. This ultimate state in stellar evolution is one stage further in compactness than the white dwarf with gravitational forces supported by degenerate nucleons. A supernovae explosion is the most likely mechanism by which this state could come into being. The density is about 10^{13} g cm^{-3} and the radius about 10 km.

(B) MECHANISMS

The fundamental problem in explaining the mechanism of the pulsations is to account for the high degree of regularity in the timing, while explaining the apparently random variations in amplitude and polarization. The good time-keeping implies that there must be a high degree of organization within the source; the intensity variations suggest that the strength of the emission is independent of the time-keeping and coupled to a relatively random process. Ostriker (1968) has pointed out that there are three processes which can give good time-keeping in stellar systems: vibration, rotation and orbital motion. In general the many theories that have been proposed to explain the pulsars involve some combination of these mechanisms with the three possible sources listed above. A few of these will be briefly considered to illustrate the complexity of the theoretical problem and hence the wide variety of explanations that have been contrived. One theory, Gold's rotating neutron star model, will be treated in more detail in the next section.

(*i*) *Vibrating sources.* Neutron stars are expected to vibrate with a characteristic period of only 1 msec. Shortly before the discovery of the pulsars, Meltzer and Thorne (1966) showed that instabilities in white dwarfs could cause vibrations with fundamental periods of 8 seconds. Subsequent attempts have been made to modify the white dwarf models in an attempt to lower this period; however it appears that white dwarfs vibrate too slowly and neutron stars too quickly to explain pulsars.

(*ii*) *Collapse.* Hoyle and Narlikar (1968) have pointed out that the time interval of one second is the characteristic time scale of super-novae collapse. They postulate that the pulsar consists of rapid oscillations between a pre-supernovae configuration and a neutron star. However this involves a tremendous dissipation of energy and the oscillations should fade rapidly.

(*iii*) *Neutron star binary system.* Since binary systems of stars rotating about their centre-of-mass are quite common, it is possible that the pulsation period could be associated with the rotation period of such a system. Because the observed periods are so short the distance between the stars must be so small that neutron stars are only the possible components. Saslaw, Faulkner and Strittmatter (1968) have proposed that if one star was a continuous source of radio emission, the other neutron star could act as a gravitational lens and beam the radiation in the direction of the observer in certain positions. However it is likely that gravitational radiation would lead to a rapid dissipation of the energy of this system.

(*iv*) *Planetary system.* A variation of the above was proposed by Burbidge and Strittmatter (1968); this system consists of a small planet in orbit about a neutron star. In the same way as the satellite Io causes radio emission from Jupiter in certain positions of its orbit, the planet here causes emission from the neutron star. However the size of the planet must be prohibitively small and the radiation mechanism is not obvious.

12.6 The rotating neutron star model

The rotating neutron star model was proposed by Gold three months after the discovery of the first pulsar [Gold (1968)]. It has since been elaborated by himself [Gold (1969)] and others [Pacini (1968)]. Its current popularity arises from the predictions that it made concerning pulsar properties which have since been confirmed. It is still only qualitative, with many features requiring further explanation.

In Gold's model the basic time-keeping mechanism is the rotation of a neutron star, with a rotation period equal to P. This rotation is also the energy source for the radio emission. Unlike white dwarfs and larger stars, neutron stars, because of their compactness, can rotate rapidly without mechanical breakdown; periods of as little as one millisecond are possible but they also can be as slow as one second.

When a star collapses to a neutron star the original magnetic field is highly compressed so that the neutron star magnetic field can be very large. Surface magnetic fields of 10^{12-13} gauss have been suggested. This strong field provides a rigid coupling between the fast rotating neutron star and its surrounding plasma, causing the latter to co-rotate with the same angular velocity, ω. The tangential velocity of the plasma at a distance r from the star centre increases linearly with r. Eventually a surface will be reached at which $\omega r = c$, the velocity of light; at this surface relativistic effects will dominate and co-rotation will cease. Plasma beyond this surface will be stationary. The radial distance will be about 10^5 times the neutron star radius and the magnetic field about 10^6 gauss. As the plasma crosses this thin surface, coherent magnetic bremsstrahlung radiation will occur. This would explain the concentration of the radiation towards long wavelengths.

To account for pulsed radiation it is necessary to imagine that the

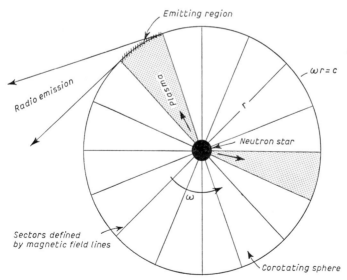

Fig. 12.3. The rotating neutron star pulses model

co-rotating plasma is divided into sectors by the magnetic field lines (Fig. 12.3). If there is an instability on the surface of the neutron star within a particular sector, plasma will be drawn out and accelerated within the sector. As it crosses the $\omega r = c$ circle the plasma radiates in the tangential direction; this cone of radiation sweeps past the observer and is detected as a pulse. This rotating light-house beacon effect is analagous to magnetic bremsstrahlung from a single particle (Fig. 2.1). For $P/T = 30$ the dimension of the surface involved in emission will be 1/30 times the circumference, i.e. about 100–1000 km.

The lifetime of the plasma in the strong magnetic field will be short so that the rapid intensity variations can be explained. The existence of a weaker instability at another point on the surface could explain the intermediate pulses. The characteristic structure of the pulsations may be the property of the instability or of the particular neutron star.

Gold has listed the following points as evidence in favour of his model:

(*i*) A change of period was predicted as a consequence of the gradual slowing down of the neutron star. This has now been observed.

(*ii*) When the shortest period was; 0·25 sec he predicted the existence of pulsars with shorter periods. The Crab pulsar was then found with a period of 33 msec.

(*iii*) Since P and T are related by geometry in this model, the quantity T/P should be approximately constant. Observations give $T/P \sim 0.07$.

(*iv*) The youngest neutron stars should have the shortest periods. The two pulsars that are associated with visible supernovae remnants have the shortest periods. The Crab Nebula is less than 10^3 years old which makes it a very young object on the astronomical time-scale.

(*v*) There is, and should be, a correlation between supernovae remnants and pulsars.

An important feature of the Gold theory is that it suggests that the pulsars are important sources of cosmic radiation. Taking the neutron star mass as $1\ M_\odot$ and radius 7 km, Gold calculated a rotational energy for the Crab pulsar of 10^{49} ergs, with the observed slowing down giving a rotational energy loss rate of 3×10^{38} ergs sec^{-1}. The energy that goes into radio emission depends on the geometry; it is probably not more than 10^{31} ergs sec^{-1}. The discrepancy between the rotational energy loss rate and the radiation energy rate can be resolved by putting the energy into relativistic particles. This would agree with the current luminosity of the Crab Nebula but initially the rate of energy loss may have been much greater, e.g. if the neutron star intially had a rotation period of 1 msec, then in its lifetime it would dissipate 10^{52} ergs, mostly into relativistic particles. This is a considerably greater emission of energy than is normally calculated for Type I supernovae. With one supernova per hundred years the Galaxy could be filled with the observed density of cosmic radiation.

With the detection of optical pulses from the Crab pulsar it is apparent that this model requires some modification. Since the optical pulses are as sharp as the radio pulses, the beaming mechanism must be equally effective at these wavelengths. It is unlikely that the optical emission could be coherent.

12.7 Applications

Whatever the nature of the pulsars the existence of sources with their peculiar properties has a number of interesting applications. Those interested in experimental methods of probing general relativity are attracted by the existence of these stable clocks far removed from the solar system. Comparison of these astronomical clocks and terrestrial clocks could indicate the perturbing influences of gravitational effects. The observed slowing down of the pulsar period could be one of these.

When the distance to the pulsars can be determined independently then dispersion measurements can be used to determine the integrated electron density along the line of sight. In the case of the Crab Nebula these measurements give $n_e < 0 \cdot 1$ cm^{-3}.

The strong polarization of the pulses can be used to probe the interstellar magnetic field by making simultaneous measurements at different frequencies; Faraday rotation in interstellar magnetic fields will cause the orientation of the vector to vary with frequency. Preliminary measurements indicate that the field is between $0 \cdot 2$ and 2μ gauss depending on direction; these values are lower than those determined by other methods.

References

Only the more basic papers are referred to here. References marked with an asterisk contain more complete reference lists.

Chapter 1

Chiu, H.-Y., *Neutrino Astrophysics*, Gordon and Breach (1965).
Payne-Gaposchkin, C., *Stars in the Making*, Harvard University Press (1952).
Schwarzschild, M., *Structure and Evolution of the Stars*, Princeton University Press (1958).

Chapter 2

Donahue, T. M., *Phys. Rev.* (1951), **84,** 972.
*Felten, J. E., Morrison, P., *Ap. J.* (1966), **146,** 686.
*Morrison, P., *Handbuch der Physik* (1961), **46,** 1.
*Shklovsky, I. S., *Cosmic Radio Waves*, Oxford University Press (1960).

Chapter 3

Colgate, S. A., *Proc. Int. Conf. Cosmic Rays, London* (1965), **1,** 112.
*Fowler, W. A., Hoyle, F., *Ap. J.* (1960), **132,** 3, *Ap. J. Suppl.* (1964), 91.
*Ginzburg, V. L., Syrovatskii, S. I., *The Origin of Cosmic Rays*, Pergamon Press (1964).
Hayakawa, S., *Brandeis Summer Institute in Theoretical Physics* (1963), **2,** 1.
*Payne-Gaposchkin, C., *The Galactic Novae*, North-Holland (1957).
Schatzman, E., *Stellar Structure*, University of Chicago (1965), **8,** 327 (Theory of Novae and Supernovae).
*Shklovsky, I. S., *Cosmic Radio Waves*, Oxford University Press (1960).
*Zwicky, F., *Stellar Structure*, University of Chicago (1965), **8,** 367 (Supernovae).

Chapter 4

Earl, J. A., *Phys. Rev. Letters* (1961), **6,** 125.
*Felten, J. E., Morrison, P., *Ap. J.* (1966), **146,** 686.

*Galbraith, K., *Extensive Air Showers*, Butterworths (1958).
*Ginzburg, V. L., Syrovatskii, S. I., *The Origin of Cosmic Rays*, Pergamon Press (1964).
Greisen, K., *Phys. Rev. Letters* (1966), **16**, 748.
Hartman, R. C., *Ap. J.* (1967), **150**, 371.
Hillas, M., *Canadian J. Phys.* (1968), **46**, 623.
Hooper, J. E., Scharff, M., *Cosmic Radiation*, Methuen (1958).
Hoyle, F., *11th Solvay Conf.* (1958), 66.
*L'Heureux, J., *Ap. J.* (1967), **148**, 399.
Meyer, P., Vogt, R., *Phys. Rev. Letters* (1961), **6**, 193.
*Shklovsky, I. S., *Cosmic Radio Waves*, Oxford University Press (1960).
Wolfendale, A. W., *Cosmic Rays*, Philosophical Library (1963).
Zatsepin, G. T., Kuzmin, V. A., *Soviet Physics, J.E.T.P. Letters* (1966), **4**, 78.

Chapter 5

*Burbidge, E. M., *Varenna Summer School*, XXXV, 'High Energy Astrophysics' (1966), 43.
Graham-Smith, F., *Radio Astronomy*, Penguin (1960).
Hewish, A., *Sci Progr.* (1965), **53**, 355.
Kraus, J. D., *Radio Astronomy*, McGraw-Hill (1966).
Matthews, T. A., Morgan, W. W., Schmidt, M., *Ap. J.* (1964), **140**, 35.
Pawsey, J. L., Bracewell, R. N., *Radio Astronomy*, Clarendon Press (1955).
*Sandage, A., *Varenna Summer School*, XXXV, 'High Energy Astrophysics' (1966), 10.
*Shklovsky, I. S., *Cosmic Radio Waves*, Oxford University Press (1960).

Chapter 6

Adgie, R., Gent, H., Slee, O. B., Frost, A., Palmer, H. P., Rowson, B., *Nature* (1965), **208**, 275.
Aizu, K., Fujimoto, Y., Hasegawa, H., Kawabata, K., Taketani, M., *Progr. Theoret. Phys. Suppl.* (1964), **31**, 35.
Arp, H. C., *Science* (1966), **151**, 1214.
*Burbidge, E. M., *Ann. Rev. Astr. and Astrop.* (1967), **5**, 399.
*Burbidge, G. R., Burbidge, E. M., *Quasi-Stellar Objects*, Freeman (1967).
Burbidge, G. R., Burbidge, E. M., Hoyle, F., *Ap. J.* (1967), **147**, 1219.
Greenstein, J. L., Matthews, T. A., *Nature* (1963), **197**, 1041.
Greenstein, J. L., Schmidt, M., *Ap. J.* (1964), **140**, 1.
Gunn, J. E., Peterson, B. A., *Ap. J.* (1965), **142**, 1633.
Hazard, C., *Conf. on Grav. Collapse*, Dallas (1963).
Heeschen, D. S., *Ap. J.* (1966), **146**, 517.
Hillier, R. R., *Nature* (1966), **212**, 1334.
Pauliny-Toth, I. K., Kellermann, K. I., *Ap. J.* (1966), **146**, 634.
Sandage, A. R., *Sky Telescope* (1961), **21**, 146.
Scheuer, P. A. G., *Nature* (1965), **207**, 963.

Schmidt, M., *Nature* (1963), **197**, 1040.
Shklovsky, I. S., *Soviet Astr. A. J.* (1962), **6**, 465.
Sholomitsky, G. B., *Intern. Bull. Variable Stars*, 83 [Com. 27, I.A.U. (1965)].
Smith, H. J., Hoffleit, D., *Nature* (1963), **198**, 650.
Strittmatter, P.A., Faulkner, J., Walmsley, M., *Nature* (1966), **212**, 1441.
Veron, P., *Nature* (1966), **211**, 724.

Chapter 7
Burbidge, G. R., Burbidge, E. M., *The Structure and Evolution of Galaxies*, Wiley (1965).
*Fowler, W. A., *Varenna Summer School*, XXXV (1966), 313.
Greenstein, J. L., Schmidt, M., *Ap. J.* (1964), **140**, 1.
Harrison, B. K., Thorne, K. S., Wakano, M., Wheeler, J. A., *Gravitational Theory and Gravitational Collapse*, University of Chicago (1965).
*Hoyle, F., Fowler, W. A., *Nature* (1963), **197**, 533; *Nature* (1967), **213**, 373.
Terrell, J., *Science* (1964), **145**, 918; *Ap. J.* (1967), **147**, 827.

Chapter 8
Bowyer, S., Bryam, E. T., Chubb, T. A., Friedman, H., *Science* (1964), **146**, 912.
Bryam, E. T., Chubb, T. A., Friedman, H., *Science* (1966), **152**, 66.
Chiu, H.-Y., *Ann. Phys. N.Y.* (1964), **26**, 364.
Clark, G. W., *Phys. Rev. Letters* (1965), **14**, 91.
Friedman, H., Bryam, E. T., Chubb, T. A., *Science* (1967), **156**, 374.
*Giacconi, R., *Varenna Summer School*, XXXV, 'High Energy Astrophysics' (1966), 73.
Giacconi, R., Gursky, H., Paolini, F., Rossi, B., *Phys. Rev. Letters* (1962), **9**, 439
*Gould, R. J., Burbidge, G. R., *Handbuch der Physik* (1967), **46/2**, 265.
Johnson, H. M., Stephenson, C. B., *Ap. J.* (1967), **146**, 602.
Morrison, P., Sartori, L., *Phys. Rev. Letters* (1965), **14**, 771.
Morton, D. C., *Ap. J.* (1964), **140**, 460.
*Oda, M., *Proc. Int. Conf. Cosmic Rays, London* (1965), **1**, 68.
Oda, M., Bradt, H., Garmire, G., Spada, G., Sreekankan, B. V., Gursky, H., Giacconi, R., Gorenstein, P., Waters, J. R., *Ap. J.* (1967), **148**, L5.

Chapter 9
Arnold, J. R., Metzger, A. E., Anderson, E. C., Van Dilla, M. A., *J. Geophys, Res.* (1962), **67**, 4878.
Chudakov, A. E., Zatsepin, V. I., Nesterova, N. M., Dadikin, V. L., *Proc. Int. Conf. Cosmic Rays, Kyoto* (1961), **3–2–9**, 106.
Clayton, D. D., Craddock, W. L., *Ap. J.* (1965), **142**, 189.
Cocconi, G., *Proc. Int. Conf. Cosmic Rays, Moscow* (1960), **2**, 309.
*Fazio, G. G., *Ann. Rev. Astr. and Astrop.* (1967), 481.
*Felten, J. E., Morrison, P., *Ap. J.* (1966), **146**, 686.

Firkowski, R., Gawin, J., Hibner, J., Wdowczyk, J., Zadwadzki, A., Maze, R., *Proc. Int. Conf. Cosmic Rays, London*, (1965), **2**, 696.

Frye, G. M., Wang, C. P., *Phys. Rev. Letters* (1967), **18**, 132.

Garmire, G., Kraushaar, W., *Space Sci. Rev.* (1965), **4**, 123.

*Ginzburg, V. L., Syrovatskii, S. I., *The Origin of Cosmic Rays*, Pergamon Press (1964).

Goldreich, P., Morrison, P., *Soviet Phys.*, *J.E.T.P.* (1964), **18**, 239.

Gould, R. J., *Phys. Rev. Letters* (1965), **15**, 511.

*Gould, R. J., Burbidge, G. R., *Handbuch der Physik* (1967), **46/2**, 265.

Gould, R. J., Schreder, G. P., *Phys. Rev. Letters* (1966), **16**, 252.

Greisen, K., in *Perspectives in Modern Physics*, Interscience (1966), 355.

Hasegawa, H., Noma, M., Suga, K., Toyoda, Y., *Proc. Int. Conf. Cosmic Rays London* (1965), **2**, 642.

Hayakawa, S., Okuda, H., Tanaka, Y., Yamamoto, Y., *Progr. Theoret. Phys Suppl.* (1964), **30**, 153.

Jelley, J. V., *Phys. Rev. Letters* (1966), **16**, 479.

Long, C. D., McBreen, B., Porter, N. A., Weekes, T. C., *Proc. Int. Conf. Cosmic Rays, London* (1965), **1**, 318.

*Morrison, P., *Nuovo Cimento* (1958), **7**, 858.

Nikishov, A. J., *Soviet Phys.*, *J.E.T.P.* (1962), **14**, 393.

Savedoff, M. P., *Nuovo Cimento* (1959), **13**, 12.

Chapter 10

Achar, C. V., Narasimham, V. S., Ramana Murthy, P. V., Creed, D. R. Pattison, J. B. M., Wolfendale, A. W., *Proc. Int. Conf. Cosmic Rays, London* (1965), **2**, 989.

Bergeson, H. E., Hilton, L. K., Keuffel, J. W., Morris, M., Parker, J. L., Stenerson, R. O., Wolfson, C. J., *Proc. Int. Conf. Cosmic Rays, London* (1965), **2**, 1048.

Bernstein, J., Feinberg, G., Ruderman, M., *Phys. Rev.* (1963), **132**, 1227.

Bethe, H. A., *Phys. Rev.* (1939), **55**, 434.

Cowan, C., Ryan, D., Buckwalter, G., *Proc. Int. Conf. Cosmic Rays, London* (1965), **2**, 1041.

Davis, R., *Phys. Rev. Letters* (1964), **12**, 303.

*Fowler, W. A., *Varenna Summer School*, XXXV, 'High Energy Astrophysics' (1966), 367.

Pontecorvo, B. (1946), U.S.A.E.C.–200–18787.

Reines, F., Crouch, M., Jenkins, T., Kropp, W., Gurr, H., Smith, G., Sellschop, J., Meyer, B., *Phys. Rev. Letters* (1963), **15**, 429.

Reines, F., Kropp, W. R., *Phys. Rev. Letters* (1964), **12**, 457.

*Ruderman, M. A., *Rep. Prog. Phys.* (1965), **61**, 411.

Chapter 11

*Dicke, R. H., Peebles, P. J. E., Roll, P. G., Wilkinson, D. T., *Ap. J.* (1965), **142**, 414.

Field, G. B., Hitchcock, J. L., *Phys. Rev. Letters* (1966), **16**, 817.

Greisen, K., *Relativistic Astrophysics Conf.*, *New York* (1967).

Howell, T. F., Shakeshaft, J. R., *Nature* (1966), **210**, 1318.

Hoyle, F., *Galaxies, Nuclei and Quasars*, Heinemann (1965).

Peebles, P. J. E., *Ap. J.* (1967), **147**, 859.

Penzias, A. A., Wilson, R. W., *Ap. J.* (1965), **142**, 419.

Stokes, R. A., Partridge, R. B., Wilkinson, D. T., *Phys. Rev. Letters* (1967), **19**, 1199.

Thaddeus, P., Clauser, J. F., *Phys. Rev. Letters* (1966), **16**, 819.

Welch, W. J., Keachie, S., Thornton, D. D., Wrixon, G., *Phys. Rev. Letters* (1967), **18**, 1068.

Wilkinson, D. T., *Phys. Rev. Letters* (1967), **19**, 1195.

Chapter 12

Burbidge, G. R., Strittmatter, P. A., *Nature* (1968), **218**, 433.

*Cameron, A. G. W., Maran, S. P., *Report on Fourth Texas Symposium on Rel. Astrophysics* (1969).

Craft, H. D., Comella, G. A., *Nature* (1968), **220**, 676.

Drake, F. D., *Fourth Texas Symposium on Rel. Astrophysics* (1968).

Gold, T., *Nature* (1968), **218**, 731; *Nature* (1969), **221**, 25.

*Hewish, A., *Scientific American* (1968), **219**, 25; *Fourth Texas Symposium on Rel. Astrophysics* (1968).

Hewish, A., Bell, S. J., Pilkington, J. D. H., Scott, P. F., Collins, R. A., *Nature* (1968), **217**, 709.

Hoyle, F., Narlikar, J., *Nature* (1968), **218**, 123.

*Maran, S. P., Cameron, A. G. W., *Physics Today* (1968), **21**, 41.

Meltzer, D. W., Thorne, K. S., *Ap. J.* (1966), **145**, 514.

Ostriker, J., *Nature* (1968), **217**, 1227.

Pacini, F., *Fourth Texas Symposium on Rel. Astrophysics* (1968).

Rickett, B. J., Lyne, A. G., *Nature* (1968), **218**, 934.

Salpeter, E. E., *Nature* (1969), **221**, 31.

Saslaw, W. C., Faulkner, J., Strittmatter, P. A., *Nature* (1968), **217**, 1222.

Subject Index

Author Index